Anoja Wickramasinghe

DEFORESTATION, WOMEN AND FORESTRY

The case of Sri Lanka

Published for the Institute for Development Research Amsterdam
by International Books (Utrecht, The Netherlands), september, 1994

CIP-DATA KONINKLIJKE BIBLIOTHEEK, DEN HAAG

Wickramasinghe, Anoja

Deforestation, women and forestry : the case of Sri Lanka
/ Anoja Wickramasinghe. – Utrecht : International Books.
– Fig., maps, tab.
Publ. in collab. with INDRA, UvA. – With ref.
ISBN 90-6224-986-8 ✔
NUGI 661
Subject headings: forestry ; Sri Lanka / women's studies.

Cover design: Tekst/Ontwerp, Amsterdam
Cover photo: ABC, Amsterdam
Printing: Bariet, Ruinen
Production: Trees Vulto DTP en Boekproduktie, Schalkwijk

Institute for Development Research Amsterdam
Plantage Muidergracht 12, 1018 TV Amsterdam, The Netherlands
tel. 31 (0)20-5255050, fax. 31 (0)20-5255040

International Books
A. Numankade 17, 3572 KP Utrecht, The Netherlands
tel. 31 (0)30-731840, fax. 31 (0)30-733614

For Ranjith and Nilanka

Contents

List of figures

List of tables

Acknowledgements

It is a pleasure to express my gratitude to the institutions and persons who have supported and encouraged me to write this book. I have gained much from women in rural areas of Sri Lanka. I was stimulated by their willingness to share their knowledge and experience. The Institute for Development Research, Amsterdam, (INDRA) offered me the financial support to concentrate on the subject while my own employer, the University of Peradeniya, Sri Lanka, extended its cooperation by releasing me. The opportunities given by both would not have been utilized successfully without information from the rural women, who spent their time in giving me the benefit of generations-old knowledge and experience. My sincere thanks are extended to those women.

I appreciate the cooperation of many of the colleagues who made it possible to bring this piece of work to a final stage. Prof. Joke Schrijvers, the Director of INDRA, for her numerous suggestions and the time spent on reading the draft. Prof. Dawn Currier, sociologist with the Department of Anthropology and Sociology, University of British Columbia, Canada, for her lively discussions, Dr. Ange Wieberdink and Ms. Heleen van den Hombergh of INDRA for providing me with a homely environment during my stay at INDRA.

At this end, I received an immense amount of support in finalising the work. My thanks are due, in particular, to Mrs. Nalani Unambuwa of the University of Peradeniya, for going through the manuscript critically, Mr. K. Rajapakse, former Assistant Conservator of the Forestry Department, for discussing the realities related to 'women's issues' and their contribution to forestry. Last not least, I must acknowledge the driving force behind this work, my husband Mr. Ranjith Wickramasinghe.

Without all the co-operation received from them I would not have shared the lessons learnt from women of rural Sri Lanka with the researchers, planners and decision makers.

Professor Anoja Wickramasinghe
Department of Geography,
University of Peradeniya,
Peradeniya, Sri Lanka

Preface

Having grown up in a Sri Lankan village, I have always been aware that indigenous forest practices are important to effective landmanagement and human subsistence. Later, as a geographer specialised in natural resource management and forest ecology, I was able to observe more systematically the indigenous forestry practices of women in the Kandyan homegardens, the forest periphery of Adams' Peak Wilderness and the farmlands in the central highlands of Sri Lanka. This gave me a more comprehensive understanding of what forests mean to people in rural areas of my country and I was particularly struck by the differences in priorities between men and women in their agricultural and forestry activities.

This book is the result of those observations and of several investigations I conducted. It attempts to analyse the forestry practices of women compared with state intervention, and to explore the extent to which state intervention has addressed the problems and needs of women who are experiencing a deepening scarcity in the supply of forest products. It also looks at the potential contribution of women to the forestry sector, using examples of women's activities in rural subsistence economies.

The first chapter outlines the problems associated with deforestation and their effects on women's lives. Later, I describe the forestry activities of rural women, the ways in which the forestry sector is structured, and the contribution of forestry activities to household survival. I go on to examine formal forestry interventions linked to state forest policies and reafforestation programmes, and make a number of policy recommendations aimed at accommodating the potential contribution of women to forestry development and enabling women to advance their status through forestry.

CHAPTER I

Introduction

In recent centuries, the rapid denudation of the earth's forest cover has aggravated numerous environmental problems which threaten human survival. The clearing of vast areas of forests for agricultural use and human settlement and their harvesting for commercial timber are widespread around the globe, while the need to conserve biodiversity and ensure environmental sustainability remains nothing more than a set of slogans.

The rate of global deforestation and the acreage under forest cover is still being debated. But it is clear that the damage already caused is irreversible. At least 120 million hectares of tropical forest were cleared between 1950-1975 in South and South-east Asia alone. Every year, about 2.5 million hectares disappear in Central America and Amazonia to make room for cattle ranching; about 1.3 million hectares in India go to commercial plantation crops, river valley projects and other activities; nearly 42,000 hectares are lost in Sri Lanka. By the year 2000, according to 1980 FAO projections, 150 million hectares of remaining tropical forest (12 percent of the total) and 76 million hectares of open tropical woodlands will be diminished (World Resources, 1986). More conservatively Vandana Shiva (1989) calculated that all tropical forests will have disappeared by the year 2000. On the other hand, reafforestation efforts will supply less then 10% of what is needed to meet Third World needs by that time (Spearns, 1978).

A quarter of the world's forest resources are in the Asia-Pacific region, where 445 million of the region's 3000 million hectares of land are covered by tropical forests. As Rao pointed out in 1992, trends in population growth and land utilization reveal that the land-human ratio in the region is progressively decreasing. The stress on regional croplands, rangelands and forests due to increasing population (estimated to reach about 3,200 million by the year 2000), he noted, represents the most significant challenge of the time. Unless this stress is relieved, he warned, we will see an unfolding environmental crisis of serious dimensions.

Timber harvesting is a continuous contributor to this situation, particularly in South-east Asia where annual deforestation increased from two million hectares in the 1970s to 4.7 million in the 1980s, according to Rao's calculations. Inappropriate land-use practices are also a giant factor, although there is no systematic information about deforestation in agricultural areas, farm lands and other scattered locations.

Important ecosystems are disappearing and hundreds of species are being lost each year. A 1986 study by the International Union for the Conservation of Nature and Natural Resources (IUCN) estimated that south of the Sahara, 65% of the original ecosystems have been disturbed, and for south Asia, 67% of the natural habitats have been lost. Estimates of this loss range from 100 species a day (Peter Reven, 1988) to 50 species a day (Wilson, 1988; Myers, 1986). However, the species extinction has been identified as a serious problem in tropical areas (i.e. Whitmore *et al* 1992).

This loss eradicates the life-support systems of rural peoples, who depend on wild species of plants, animals and fish for food, timber, medicinal herbs, fibre, and fuelwood. The livelihoods, nutrition, health, protection and local trade (Abramovitz, 1991) of local people are based on forest resources. Diversity in biological resources has enabled them to assure diversity in family diet, availability of food commodities and medicinal products.

With the disappearance of forests, seasonal droughts have extended into vast areas, creating seasonal deserts. According to the World Commission on Environment and Development (Bruntland, 1987), six million hectares of productive lands are being turned into worthless desert each year. Globally, some 456 million people are starving or suffering from malnourishment as a result. The drying out of perennial streams and springs has not been treated as a serious issue, although the resulting scarcity of fuelwood is now being given attention. According to the FAO, by the end of the century more than 200 million people in developing countries will suffer from an acute scarcity of fuelwood. When fuelwood is beyond the reach of poor households, these households cook fewer meals, eat less nutritious food, and face serious malnutrition (Snyder, 1990; Wickramasinghe, 1990a). Degradation of many of the remaining resources as a result of over-utilization and destruction by local people who have nowhere else to obtain their subsistence requirements will be far more serious than at present.

These developments have major implications, particularly for women who use forest resources for their families' survival needs. In most cases, the conversion of forest areas to other uses has resulted in changes in land rights. Communal rights over land shared by people in harmony with nature is giving way to private ownership. This restricts women's access to land owned by their families or people they know. The tradition of dividing family farmland among male children makes this situation worse. Farming, the productive task in which most rural women are engaged, has become an unpaid service rendered for the well-being of the family. Although women share the responsibility of crop production, they have not received equal treatment in land development policies in Sri Lanka. Moreover, in modern commercialised agriculture, they are not treated equally because of weaknesses in policy formulation and an absence of any effort to recognise their rights to land. Since legal land ownership is a condition of eligibility to facilities and services, they are then denied these as well.

The result of this complex of factors is an increasing fragmentation of farmland, which forces women from poor families to work for wages outside the farms. In addition, the supremacy given to technical and scientific knowledge over traditional practices has denied rural women a role in decision-making on land matters and discredited their expertise. They are thus unable to contribute to a sector in which they have a great deal of experience. As Rao stressed in 1987, to restore the balance between the needs and the availability of forest sources, it is essential that women be promoted in forestry. The Food and Agriculture Organisation of the United Nations (FAO), by establishing its action programme on Forestry for Local Community Development in the late 1970s, stressed the importance of trees in rural economies. However, in the recent past, although the FAO has published a number of studies on women and forestry, its Tropical Forestry Action Plan (1985) has not given adequate attention to women.

CHAPTER 2

Women and forestry: views and perceptions

As a development sector, forestry is male dominated. Decisions made at the level of policy planning and regional implementation are accepted as infallible. Women, who form the lowest strata of society, must implement decisions which ignore their own urgent needs.

Women's use of forest resources

Rural women in developing countries use forests as a provider of items essential for household survival. Food for the family, fuelwood for domestic cooking, fodder for domestic animals, fibre and raw materials for making household utensils, pharmaceutical products to treat ailments and prevent the spread of disease – all these are acquired from the forest. For centuries, these resources were widely available and women used them freely. In the process, women accumulated an immense store of knowledge about forest resources, where resources were available, how they could most efficiently be utilised and conserved for future generations. One visit to the forest would serve multiple purposes, as generations of knowledge would be drawn upon. The food security of rural populations has therefore always been directly dependent on forests and farm trees.

In 1987, Falconer argued that although forest foods were not dietary staples, they were important supplements, which increased the overall diversity and nutritional quality of the rural diet. Processed and stored forest food products were essential to a year-round food supply. However, more recent analysis of village and forest tree-use practices of rural households in Asia reveals that, in fact, subsistence items also come from the farm trees and the forest. (Wickramasinghe, 1993a). Data on farm and village forestry practices in Bangladesh, Indonesia, Thailand, Philippines, Nepal and Sri Lanka reveals that among 10 categories of products gathered from forests, the ones most widely used are food, fuelwood and fodder (see Figure 1).

Figure 1 *Pattern of using tree products by rural households in Asia (%)*

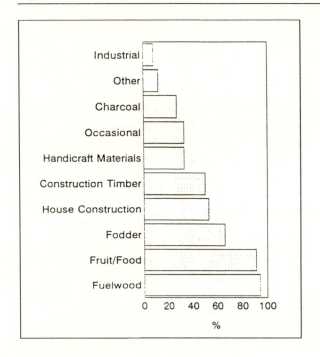

Source Wickramasinghe, A. (1993b)

Nearly 86% of products are for household use. On-farm sources provide 62% of the tree products (Raintree, 1992). 94% of households use village sources for obtaining fuelwood; 91% and 66% for food products and fodder respectively. The pattern of percentage distribution of women as gatherers among all products reveals that these three products are their priority. As Rodda observed in 1991, 'Traditionally, women have gathered products which have provided them with the basic three 'Fs' of fuel, food and fodder, and for a variety of other uses.' Nearly 73% of women in Asia concentrate on obtaining food, fuelwood, and fodder products.

Figure 2 *Bar-graph showing the % distribution of women as collectors among 5 categories of products in 6 countries in Asia.*

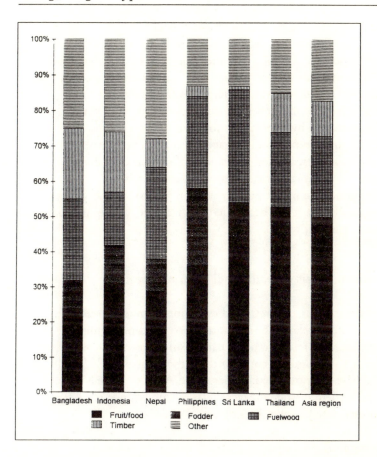

Source Wickramasinghe, A. (1993c)

If one breaks down these figures on a country basis, one gets the following: 55% in Bangladesh, 57% in Indonesia, 64% in Nepal, 84% in The Philippines, 86% in Sri Lanka and 74% in Thailand (see Figure 2). This pattern implies that women give domestic needs higher priorities because of their conventional roles of wives, mothers and homemakers. Men are mainly engaged as collectors of timber, construction materials, and products for other uses.

Due to the recent recession of forests and the liquidation of women's rights to them, such multiple uses of forests are now in decline. This decline has fragmented forest-based cultures in Asia, Latin America and Africa, where the women have transferred their knowledge into farm-forestry practices. The traditional pattern is in direct contrast with the way modern forestry intervention is organised. Even aged forest monocultures of teak, eucalyptus and pine are managed by state sectors for their timber. Meanwhile, women in rural areas are still engaged in managing homegardens of multiple structures, species, and compositions for their diverse products (Wickramasinghe, 1991a). This contrast is well-demonstrated by the priority women give to the use-values of *Artocarpus heterophylus* – designated the 'rice tree' of the rural poor (Wickramasinghe, 1991b).

The cultural basis of traditional form of forestry

With the shrinking of forests and extension of modern scientific thinking, rituals related to trees and forests are diminishing. Nevertheless, the resource-poor, particularly those who are in close proximity to forests, still consider them the source of life. The Chipko Movement in the Himalayas and Appico in the Northern Karanataka demonstrate the close relationship women feel with trees and forests. The Chipko slogan goes:

"What forests bring us;
soil, water and clean air;
soil, water and clean air;
the basis of our life."

Here are the words of a woman from Kudawa village, Adam's Peak Wilderness, who was interviewed in December 1991 bearing a headload of fuelwood, a bundle of greens wrapped in a palm leaf and a bunch of fruit of Atamba (*Mangifera zeylanica*) tied to her waist:
"It is an asset gifted to us by nature. Almost every day I collect a few varieties of seeds, mushrooms, fruits, green wild leaves and, of

course, a bundle of deadwood for cooking. These forest foods are of special tastes and we cannot find them elsewhere. See the medicinal herbs that I have collected today? We don't have to experience hunger. I shall prepare leaves in porridge, and dry all the stiff parts to use in medicine or as beverage... "

Whether commonly eaten, sought after as a rare delicacy, or consumed after complicated processing in times of scarcity, almost every plant form found in the forest is a source of food for people who live nearby (de Beer *et al*, 1989). Herbs, plants and leaves of wild woody species are collected as vegetables, porridges, beverages, sauces, condiments, medicines and flavourings. Crystal sugar and treacle, made out of the floresen sap of Kitul palm (*Caryota urens*), are popular among peripheral forest dwellers. Roots, fungi, rhizomes, nuts, seeds and tubers are their favourite foods. Fruits are either cooked or eaten raw. Edible and medicinal oils are extracted out of nuts, kernels, and seeds. In most cases, the products garnered from the forest include items for immediate home consumption, raw materials and marketable commodities (see Table 1).

The relationship with the forest is the basis for the entire material and social culture of these people. Forests offer potential, while indigenous knowledge, technology, needs, priorities and gender determine the tasks and roles of men and women in handling these raw materials. Products with cash potential are the primary concerns of men, while women are mainly concerned with products for family use. For women, only the excess goes to the market.

In the forest men, women and children are the gatherers, explorers, users, and also the guardians. When the forest products are brought into their dwellings, a multitude of tasks emerge (see Figure 3). Fuelwood, fruits, other foods, medicinal herbs, binding materials, honey and resins are used at their households, while the excess is processed either for the market or for the off-season. Treating ailments and preventing disease using medicinal products of the forests is traditional. Tasks related to building dwellings from forest timber, palm leaves, bamboo, rattan, and binding materials are multiple.

Swaminathan's experience with tribals in Rajastan has made it clear.

'There is nothing we can teach the tribals about trees,' he wrote in 1982. 'They know that rain falls heaviest where there are more trees.

They know that soil erosion and the run-off increase with deforestation. They know the ecological changes that happen with hills being denuded of forests, how the soils become dry, sterile, hard and unyielding. The forests are intricately woven by women into their culture, religion, songs, dance and rituals.'

Figure 3 *Multiple roles and tasks in the management of forest resources for survival maintenance.*

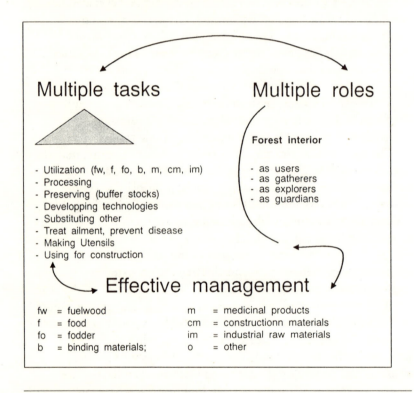

Source Wickramasinghe, A. (1993e)

Tasks related to forest management are performed by men and women (Wickramasinghe, 1993e). The pattern of men's and women's management in gathering and processing reflects differences in their priorities, experience and knowledge (see Figure 4).

Figure 4 *Gender specific division of tasks related to forest management*

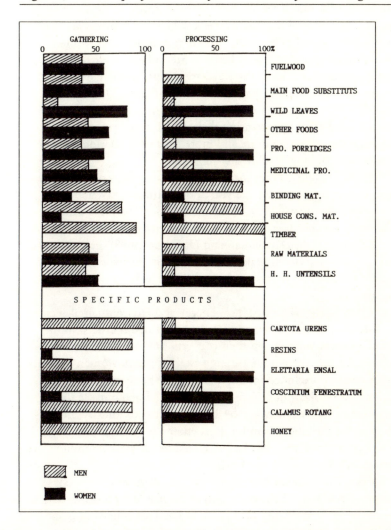

Source Wickramasinghe, A. (1993e)

Women as victims of deforestation

At international forums, deforestation is treated as a serious problem, but importance is attached to the consequences for the global environment and not to these women dependent on forest resources for subsistence.

In Sri Lanka, out of the 23 types of forest products gathered by peripheral dwellers of the Adams' Peak wilderness, women are the primary gatherers of 21 products. About 31 products that women gathered in one area have been listed in Sierra Leone (Hoskins, 1979 a). In Bushman societies of the Kalahari Desert, women are major food providers through gathering in the woodlands (Draper, 1975). Even with forests getting scarce, women continue to carry on conservational tasks by increasing their working hours. They walk long distances, spending more and more time on gathering fuelwood for domestic cooking than on anything else.

It is now recognised that there is a developing fuelwood crisis of tremendous magnitude. By the year 2000, estimates say, about 2.4 billion rural people will be using fuelwood faster than it is being replenished (Eckholm *et al*, 1984). The evidence provided by Agarwal (1986); Shiva (1989); Dankelman and Davidson (1988); Eckholm *et al* (1984); Postel *et al* (1988) makes it clear that the fuelwood scarcity problem needs to be addressed through forestry development programmes.

As Cecelski (1992) argues, 'how energy and environmental-sector problems and objectives are perceived and defined is crucial in the light of whether women and their concerns are seen as essential or marginal. In the conventional energy establishment, women have not been necessarily excluded intentionally or their energy-related activities overlooked; they have simply been defined outside the energy sector". As in many of the aspects related to forestry, in the energy sector too, the reason for most energy project failures is not the lack of technologies, but factors traced to the socio-economic and cultural realm.' (Tinker, 1987)

But the fuelwood scarcity is only one problem. Home-based enterprises also require the collection of diverse other forest products – fruits, leaves, gums and resins (FAO, 1991). This collection is an activity in which women are heavily involved. Household food security, ensured through the use of forest foods, is also threatened.

Although the viability of managing forests using non-timber forest products is discussed, (Panayotou *et al* (1992), strategic measures to restore such productive functions are given less importance than those connected with the environmental and fuelwood problems caused by deforestation.

Table 1 *The Use Patterns of Forest Products*

Use of products	% households in each category	Use Patterns	
		household use	markets
Fuelwood	100	***	-
Substitutes for staple diet	62	***	*
Wild leaves	84	***	*
Fruits	49	***	*
Other Foods (nuts,seeds, fungi,etc.)	82	***	*
Products for Porridges	69	***	*
Medicinal Products	71	***	**
Binding materials	80	***	-
House construction materials	80	***	-
Timber	73	*	*
Products for Home Based Work	64	**	***
Raw materials for Utensils	65	**	***
Other specific species/products			
Caryota urens[Kitul]	36	**	***
Raisins	22	-	***
Elettaria ensal[Wal-ensal]	34	**	***
Coscinium fenestratum[Weinwel]	14	**	***
Calamus rotang[Wewel]	16	*	***
Honey	14	*	***
Garcinia cambogia[Goraka]	29	**	***
Cinnamomium multiflorum[Walkurundu]	28	**	***

*** = Primary; ** = Secondary, * = Occasional

Source Field information (1991)

Efforts to combat deforestation

According to the FAO's Tropical Forestry Action Plan, the underlying causes of deforestation and land degradation are: poverty, inequitable land distribution, low agricultural productivity, poor land-use policies, inappropriate development, weak institutions and rapid population growth. Clearly, increasing population and the resulting demand for land is one cause under which many others can be subsumed. Although forestry has been recognised as the solution to global warming, desertification, atmospheric pollution and land degradation – all of which threaten global life – appropriate strategies to restore forests have not emerged. Progress in forestry is extremely slow, in spite of huge investments, numerous scientific investigations, technological and institutional improvements, formalisation and systematisation. From 1950-1975, at least 120 million hectares of tropical forests were cleared in South and South-east Asia in spite of strict guardianship, private ownership and legislation.

The reasons for this failure lie in the concepts of forestry which have been employed in dealing with the problem. When reafforestation became a 'development sector'- aimed at restoring the wealth lost through deforestation – nature's control over natural processes was replaced by scientific knowledge. Forestry sectors came under the guardianship of technically-trained foresters and peoples' participation was simply regarded as support for state programmes. However, socio- economic and anthropological studies conducted in many regions show that, unless subsistence uses are guaranteed peoples' participation will not be effective. Besides the problems related to acceptability and adoptability of forestry programmes introduced through experimentation and testing, their execution in well-established forestry-related culture have been dogged with problems. Because programmes excluded the art of forestry practised by people, particularly women, whose forest-management skills have evolved over generations.

Current-day forestry is not sustainable because it ignores the indigenous knowledge which guided rural women to employ the best methods of preservation yet known to mankind. Where legislation has tried to alienate people from their traditional rights to forests, acceptance of the harmonious nature of gathering forest products for subsistence still made it possible for women to creep into the forests as gatherers (Wickramasinghe, 1993d).

As Raintree observed in 1991:

'If there is one lesson to be drawn from the past decade of experience in tree-planting projects, it is that there is no substitute for local experience. The indigenous knowledge and decision making strategies of local people represent age-old accumulated local experience which is, in the first instance, the best of project guidance.'

Shiva *et al* (1985) support this. 'Forest settlements produce the best scientific research and cultural writings. Yet it is not practically possible to explore them without having sufficient experience. However, with deforestation, strict legislation was enforced, eliminating people out of the forests for the sake of conservation and forest-based living has been detached. Still, for many of the indigenous people, living closer to the forest has been a privilege. In most cases it has been a place for meditation – psychological relaxation'. The global situation reveals that for about 200 million people, the forest is their only home (Dankelman, *et al*, 1988).

Hopes of benefiting from nature's gifts are vanishing. The accumulated knowledge regarding the ecological niches in which specific tree species flourish, and their utilization practices have not been translated into forestry practices. The adoption of trees into household lands is not new. But inadequate understanding of the links between trees and human survival has created a gap between the forestry based on science and the forestry based on the art of living.

A changing concept of forestry

There is now a growing realisation, among nations and institutions, that forestry issues cannot be separated from people's needs. This means not only solving the problems which are deepening with increasing population, but also recognising that forestry development cannot be achieved without the commitment of people. The paradigm under which forestry operates has thus been expanded to include 'social forestry'. But this concept has an inherent flaw, as Leslie pointed out in 1987.

'Forestry, after all, originated centuries ago, in the needs of specific services which forests, properly managed, can provide for as long as those needs persist. In that sense, all forestry is social. It is a little confusing, therefore, to coin the term social forestry as a special form of forestry.'

He argued that forestry which is about people is broader than forestry simply as a supplier of commercial, or even public, goods and services. For the people, who are closer to the local resource base including its flora and fauna, there has been no specific profession related to the management of those resources. Technologies of forest management have simply been part of their life-styles since time immemorial.

At the United Nations Conference on Environment and Development (UNCED) in Rio de Janeiro in June 1992, it was recognised that forest management for multiple uses is an effective, long-term method of combating deforestation. Management plans, the conference concluded, should involve all special interest groups, including indigenous people, rural women and ethnic minorities. Since 1978, the FAO has been pursuing this in its programmes by adopting the theme, 'Forestry for Local Community Development'.

This concern has created a change in the concept of forestry (Rao 1990). A socio-economic dimension has been added to the discipline. This change calls for a new breed of foresters who are not only well versed in science, but are also good communicators and motivators. It has been recognised that people's needs have to be taken seriously by redesigning forestry and encouraging people's participation. Most contemporary definitions now focus on the control and management of forest resources by the rural people who use them for domestic purposes and as an integral part of subsistence and peasant farming systems (Fisher and Gilmour, 1990). This expanded definition implies that the people who carry out the tasks of forest management should be placed at the centre of community forestry development.

However, the terms 'people' or 'community' are interpreted vaguely, so it has not been possible to make allowances for differences in communities related to their problems, priorities, gender and socio- economics. Even when the needs and priorities of a concerned group is identified as the basis for the effectiveness of forestry programmes, it is the experts who define these needs. When individual households are consulted, women's needs are defined by men. So community forestry programmes in the developing world have become limited by traditions of male leadership and power. This has marginalized the women who actually practice forestry. Programmes have focused on timber production alone; practicalities related to the multiple uses and sustainable management of the forests have been de-prioritised. Women's knowledge of the trees and plants that help to meet local needs has been ignored.

The role women can play in making forestry successful

It is obvious that there is a widening gap between forestry needs and supply. Women's contribution to forest management and conservation is vital, since they gather forest products in small quantities without destroying the resources. According to Fortman (1986), they have more intimate knowledge of local forest resources than men and foresters from other locations. She notes that:

'Local women can differentiate among species that provide long-lasting low heat, a quick high heat, or those which smoke and so on. Where wood is the main energy source for cooking, a knowledge of such characteristics is important for planning. Without the benefit of women's knowledge, foresters could produce a plantation of quick-growing wood that would not meet local needs.'

This example also applies to other tree products. The sustainable use of forest and tree resources depends on women's participation on the following grounds:

i Women's multiple uses of forests is a means of combating deforestation and conserving biodiversity;
ii Women's interest in securing the subsistence needs of their families lead them to be protective of forest resources;
iii The amount of time and energy spent on walking distances to sources of forest products gives women an interest in promoting forestry;
iv As equal partners in the development process, women should have equal opportunities in the process of forestry development.

During events connected with the UN Decade for Women (1975-1985), it was recognised that ways should be found to accommodate women's knowledge in reformulating local, national regional and also global strategies. In 1993, the Secretariat Note of the 15th Session of the Asia-Pacific Forestry Commission identified a number of key areas where change was needed to ensure equal participation by men and women. These concerned: access to natural resources, particularly tenurial rights to land and trees; structures of forestry institutions; community organizations and intersectoral linkages between forestry and other sectors.

Later in this book we will examine all these areas to see whether these policy changes have filtered down to the practical arena or have remained at the level of international rhetoric.

Development, sustainable development and forestry

In its conventional form, forestry has been primarily focused on protecting the existing forests on the one hand and increasing timber production through establishing plantations to cater for the increasing demands of the industrial sector on the other. Originally, there were no links between the paradigm of forestry and a development model which emphasises improving the standard of living of the poor.

An integration of these two paradigms took place in the late 1970s, with the adoption of development thinking into forestry and the re-formulation of forestry for the needs of the people. Over the 20 years since the United Nations Conference on the Human Environment in 1972, questions have been raised about lack of adequate development in Third World countries, continuing environmental deterioration and deforestation. The FAO, the leading international organization concerned with the forestry sector in the Asia-Pacific region, created a landmark with the publication of 'Forestry for Local Community Development' (FAO, 1978). This publication called for a great change in the forestry sector itself and integrated development thinking into forestry. It has placed enormous demands on the forestry professions to grow beyond their traditional role as guardians of public sector resources toward a larger, more diverse and socially-responsive role in extension and community development, working increasingly with private individuals and small groups (Raintree, 1991).

In the eighties, an awareness of the seriousness of the environmental crisis seized global and national attention. Recognition of increasing ecological stresses, global warming, atmospheric pollution and desertification have drawn the attention of nations to concentrate heavily on strategic global solutions. This attention has led to the 'greening' of development thinking (Harrison, 1987; Conroy and Litvinoff, 1988 and Adams, 1990) and the birth of the concept of sustainable development. The links between poverty and environmental degradation began to be seen as reciprocal, poverty being a major cause as well as an effect of global environmental problems. This idea became the central concept in the World Conservation Strategy (IUCN, 1980) and the report of the World Commission on Environment and Development, published in 1987 under the title *Our Common Future* and popularly known as the Brundtland report.

'There has been a growing realisation in national governments and multilateral institutions,' it states, 'that it is impossible to separate economic development issues from environmental issues; many forms of development erode the environmental resources upon which they must be based, and environmental degradation can undermine economic development.'

According to this document, sustainable development is 'development which meets the needs of the present without compromising the ability of future generation to meet their own needs'. International agencies, such as the United Nations Environmental Programme (UNEP) and the International Union for Conservation of Nature and Natural Resources (IUCN), have encouraged a number of programmes based on these ideas.

But, as Bhasi *et al* argued in 1991, besides economic and ecological dimensions, social, and political conditions are crucial to issues of development. As Adams pointed out in 1990, the crucial factor is not the way the environment is managed, but who has the power to decide how it is managed. This focuses on the capacity of the poor to exist on their own terms. Sustainable development is thus the beginning of the process, not the end.

It is clear that thinkers like Bruntland have not gone into these issues in sufficient depth. She has given little attention, for example, to the fact that women's potential engagement in managing the resources while utilising them is crucial to sustainable development. Where forestry is concerned, the greening of development has shifted the focus towards problems such as soil erosion, increasing droughts and floods, disrupted nutrient cycles, reduced water resources and soil productivity.

'Sustainable forest development' became a major focus of the United Nations Conference on Environment and Development in 1992. Its report, Agenda 21, insists that forest resources are essential to both development and the preservation of the global environment. Their use helps to generate employment, alleviate poverty and provides a valuable range of products. Concepts that have been integrated at this gathering of global nations are not only related to forest production, but also to the integration of forest utilization into forest management. Agenda 21 has called for the participation of people to support and develop the multiple ecological, economic, social and cultural roles and function of trees, forest and forest lands, including interest groups, women and indigenous peoples. The first priority is to 'secure the multiple roles of trees, forests and forest lands by strengthening national institutions and capabilities to formulate and implement effective policies, plans, programmes and projects relevant to forest issues'. In addition, the importance of social, economic and ecological values of trees, forests and forest lands have been recognised.

However, our knowledge and information about the value of forests and uses of forest products is sketchy. Assessments of forest products made in economic terms overlook the cultural base that overwhelmingly affects the users, and have been a constraint to policy formulation. In theory, the concept of sustainable forestry development is attractive for forestry policies. However, in practical terms, when programmes are executed, the users of forests, particularly women, are less recognised. Therefore, modern forestry becomes a group of technocratic recipes without the broad cultural understanding that is essential to effective forest management.

Whatever the concepts we may adopt in forestry development, it is clear that many of the existing forestry efforts are short-term solutions which do not serve the interests of sustainable development and the needs of future generations. To give an example: it has been fallaciously argued that fast-growing exotics are the only species that could be used to solve critical problems. Yet most of the fast-growing forest stocks suffer from either lack of regeneration or extremely slow growth processes. Once the stocks are harvested for industry, lands become more denuded than ever before. It is clear that this strategy suits the needs of wealthy entrepreneurs but not those of local inhabitants, far less future generations. No sustainable development can be achieved when programmes are designed by external experts.

Equity in forestry development

General concerns about equity and equal opportunities for women emerged in the seventies and eighties. In Asia, it was noted that women, particularly in the rural areas, work longer hours than their male partners but received less remuneration and recognition. Attempts have been made to redress this situation through the advancement of women. This advancement can be achieved both by fairer treatment and by giving more recognition to their substantial contribution to society, which has too long been underestimated. Thus, as the Directorate-General for International Co-operation noted in 1990, while an improvement of the position and status of women is totally valid as an emancipatory end in itself, utilization of women's potential contributions is also an efficient means to improve the quality of development as a whole.

With regard to forestry development, as in case of any other sectoral development, women can claim a place as potential contributors which can in turn contribute to the advancement of their socio-economic status in general. Here, their particular management skills are in great demand. Equity for women in forestry is therefore a claim both for success as well as survival. But this requires policy reformulations as well as strategic transitions (Wickramasinghe, 1992a). The conservative ideas that consider forestry as a masculine profession need changing. The priority placed on guardianship rather than on extension is a further limitation.

Foresters are not trained in social issues and extension. Women's efficiency in such activities is deliberately overlooked. Casual employment in raising nurseries are considered enough opportunities for them, but equal rights permitting the harvesting of trees which they have nurtured must be ensured through legislation. It is an injustice to expect women to tend trees for the sake of by-products, but to refuse them permits to harvest timber because the land ownership is on the name of their husband, for example. Legal procedures involved in issuing permits to cut species like jack trees vests rights in the legal owner. This means that women can only get returns from by-products, while men can sell timber for cash. This, in turn, prevents women from giving priority to timber. Equity in forestry implies not just the positioning of women on the professional ladder but acknowledging their rights over trees and lands.

Domestic energy problems

Woodfuel is a vital source of energy in most developing nations, including Benin, the Central African Republic, Chad, Ethiopia, Rwanda, Uganda, Upper Volta and Nepal. Fuelwood is of central importance to women, as the non-availability of fuelwood directly affects the family food intake. When there are severe food scarcities, although free food subsidies may be delivered, there are no such systems regarding fuelwood. Thus, acute scarcity of fuelwood could lead to low food intake, even if raw food commodities are made available. At present, as has been revealed by the FAO in 1981 (see Table 2), that the relative availability of woodfuel varies by several times above the estimated need in the high forest areas of the world to much below the needs in the arid and sub-arid zones.

Table 2 *Fuelwood use by ecological regions 1980*
(in m³ per capita per annum)

Region	Needs	Availability
Africa (South of Sahara)		
Arid & semi arid areas	0.05	0.05 to 0.01
Mountainous areas	1.4 to 1.9	0.5 to 0.7
Savanna areas (a)	1.0 to 1.5	0.8 to 0.9
Savanna areas (b)	1.8 to 2.1	
High forest areas (a)	1.2 to 1.7	1.8 to 2.0
High forest areas (b)	5.0 to 10.0	
Asia		
Mountainous areas	1.3 to 1.8	0.2 to 0.3
Indo-Gangetic Plains	0.2 to 0.7	0.15 to 0.25
Low land areas of S.E. Asia	0.3 to 0.9	0.2 to 0.3
High forest areas	0.9 to 1.3	1.0 to 6.0
Latin America		
Andean Plateau	0.95 to 1.6	0.2 to 0.4
Arid areas	0.6 to 0.9	0.1 to 0.3
Semi-arid areas	0.7 to 1.2	0.6 to 1.0
Sub-tropical & temperate areas	0.5 to 1.2	1.9 to 2.3
Abundant forest areas	0.5 to 1.2	2.5 to 10.0

Source FAO (1981): Map of fuelwood situation in Developing Countries, Explanatory Note, FAO, Rome.

According to this study, over 100 million people already face an acute fuelwood scarcity, and 1-3 billion people (39% of the total population of the developing countries) live in deficit areas and thus face a deepening crisis. The deficit situations are expected to lead to acute scarcity and, where availability is now high, scarcities are projected. In all, by the end of century, some 150 million rural dwellers will suffer acute scarcity and some 1-8 billion will be overcutting available forest resources (Eckholm *et al*, 1984). The failure to combat this situation at present is not merely due to a limitation of macro-level efforts, but also to the bias against stimulating people, particularly women, to act in their own country situations.

Table 3 *Distance Travelled and Time Spent in Collecting Fuelwood in Sri Lanka*

Area	Distance per trip-Km	Time Hours/week
dry zone		
I. Kelegama	1.5 - 2.5	13
2. Thorowa	0.5 - 1.5	11
intermediate zone		
3. Hapuwala	1.5 - 3.0	15
4. Bambarabedda	1.5 - 3.5	14
5. Madiwaka	1.5 - 2.5	13
6. Hulandawa	0.5 - 1.0	9
7. Narampanawa	0.5 - 1.5	12
wet zone		
8. Millewa	0.5 - 2.0	11
9. Meegoda	0.5 - 1.0	10

Source Author's field work, 1989.

The highest consumption recorded in Hulandawa, where nearly 55% of the fuelwood is obtained from natural forests, is attributed to the excessively high weight of the wood they burn. In areas where women either walk long distances in search of wood or have no prominent sources, care is taken to consume less wood by using other substitutes like coconut husks, shells and fronds which do not fall into the category of wood. However, where there is heavy consumption, the time spent on gathering and the distance that women walk is high. Bambaradedda, Hapuwala and Mediwaka, which are examples of low-consuming areas, demonstrate the ways in which time and distance have a bearing on consumption (see Table 3).

The amount of wood carried in a headload is related to the distance that has to be travelled. More and more trips are made on gathering fuelwood when women carry them in small bundles. Where women walk long distances, they tend to carry bundles weighing 28-34 kg of wood per trip, in spite of the difficulties incurred. In addition, they turn to their village resource-bases, including small patches of riverine vegetation, common areas, homegardens, hedges and isolated trees in their farmlands (see Table 4).

Table 4 *The Changes in the Sources of Fuelwood for Domestic Use in the Study Area*

Source	1950	1988
Natural forest	70	0
Common / reserve land	10	58
Own homegarden / hedges	10	40
Farmland	10	2
Total	100	100

Source Author's field work (1990).

The inability to use forest resources for fuelwood has created pressure on other freely-available sources like common land and riverine reserve land. The increase in percentage share of own sources between 1950 and 1988 was not due to the enrichment of their own sources with fuelwood species, but to the utilization of crop residue as fuel. Even small sticks which produce a low heat and coconut fronds, shells and husks have increased in use over the past few decades. This results in the reduction in organic residues used to enrich soil.

In spite of these trends, foresters in Sri Lanka have argued that there is no felt need for fuelwood. This is true when compared with the critical situation in Nepal, Bangladesh and India. There are no records to conclude that people have starved due to lack of fuelwood. But the question which is never raised is whether fuelwood has been provided with great difficulty in terms of women's energy and time. If we take it for granted that fuelwood is available because it comes to the kitchen hearth without spending cash, then we are overlooking the strain on over 90% of the rural women who take on the drudgery of collecting fuelwood and cooking meals to feed us. Moreover, burning twigs, wood of low calorific values and coconut husks have made cooking hours longer and more energy consuming.

The connection between the fuelwood crisis and women's work is direct (Celeski, 1985b, 1987 and Saxema, 1987). Yet, it is difficult to assess the severity of the problem due to spatial variations and the complexity of influential factors. However, studies have shown a substantial increase in time and energy spent on gathering fuelwood. It has quadrupled in parts of Sudan, according to Digerness (1977). In Nepal, gathering of fuelwood and fodder is now reckoned to take a whole day, instead of the hour or two it took a generation ago (Eckholm, 1975). As Agarwal (1986) shows, in parts of Bihar in India, about 7-8 years ago, women of poor rural households could get fuel within a distance of 1.5 to 2 km. Currently, they have to walk 8-10 kilometres every day. Similarly, in Gujarat, the severity of this problem has resulted in the use of roots, bushes and weeds which do not provide continuous heat (Nagbrahman *et al*, 1983). All this research needs to be taken seriously, and the effects on the household survival, and resources must be analysed.

Cooking time has increased and cooking has become an overwhelming burden. Women are being forced to abandon productive tasks and domestic activities to fulfil this survival need. In Sri Lanka, this effect has not been taken into consideration. Foster (1986) reveals that, in Perú, up to 10% of women spent virtually all their time on collecting fuelwood, while in Gambia, they do so from midday to nightfall every day. The extremely high cost of wood scarcity is pointed out by Postel and Heise (1988). In rural parts of the Himalayas and African Sahel, women and children spend 100 to 300 days a year on gathering fuelwood.

Soon, gathering fuelwood will be beyond women's capacity to spend time, on top of all other time-consuming tasks like washing, cooking, producing crops, processing of crop products and caring for children. Already, they are making adjustments such as reducing consumption and compromising traditional food habits. Sometimes meals are foregone. A study by Kumar and Hotchkiss in 1988 examined women's time allocation and its effects on agricultural output, food consumption and nutritional levels. They found that families have had to reduce the number of cooked meals in Bangladesh (Hughart, 1979) and the Sahel in Africa (Floor, 1977). In Guatemala (Hoskins, 1979b), the fuelwood scarcity is forcing families to abandon their traditional time- and energy-consuming diets, ignoring the consequences for nutrition. Postel and Heise (1988) have described situations in which boiling of water has become an unaffordable luxury and quick-cooking tubers and cereals have replaced slower-cooked food due to fuelwood scarcity. In Sri Lanka, women now avoid the par-boiling of rice (Wickramasinghe, 1990a).

Food security has been affected since, under severe scarcity, preservation of excess products by smoke-drying has become difficult. Processing of Artocarpus seeds and flakes for off-season consumption, for instance, has been affected by the fuelwood scarcity. The solution which has been tried – improved wood stoves – only replaces isolated parts of the kitchen hearth's functions and has not made a significant contribution.

These compromises have a great impact on food habits of the people in the tropics, where, traditionally, longer-cooked hot meals are appreciated. Poulsen (1978) shows that none of the principal food crops of the tropics are palatable in a raw state. Apart from reducing the number of cooked meals, the shortage of fuelwood eliminates the use of food varieties produced in farmyards and home gardens. Health is also affected, since boiling water for drinking as a preventative measure becomes difficult.

In this way, the fuelwood scarcity is not an isolated factor, to be looked at simply in terms of the cost in women's energy and time. Increasing food scarcity and reduced diversity of family diets, reduced women's participation in crop production and deterioration of their own health and that of their families are some other aspects which need to be acknowledged. The changing of food habits and avoidance of long-cooking processes mean an under-utilisation of some food-stuffs available locally. In many cases, the extra time for fuelwood collection is earned by sacrificing women's leisure and self-care.

Growing concern about the impact of the changing environment and depletion of biomass fuel on rural women has drawn attention to the health consequences of the situation. The results of investigations conducted in various parts of the world highlight some global problems related to women and domestic energy. Women who spend longer periods in smoky kitchens suffer direct health consequences (WHO, 1984).

Yet the effects of fuelwood shortages on women's health, strength and working capacities are not emphasised. At the ILO meeting in Geneva in 1985, the health hazards of carrying extremely heavy loads and of exposure to smoke in cooking were discussed. Throughout the world, women are carrying headloads of up to 35 kilograms of fuel as

35

far as 10 kilometres (Dankelman *et al,* 1988). This exceeds the maximum weight permissible by law in many countries, which prohibits the manual carrying of fuelwood loads heavier than 20 kilograms by women (ILO, 1966). Such heavy burdens damage the spine and interfere with childbearing. The back-breaking work of collecting, cutting and carrying wood, aggravated by poor nutrition, undermines women's health still further, and the longer they have to walk, the more they are affected (Dankelman and Davidson, 1988). Health problems, such as eye and respiratory diseases, also arise as a result of lengthy exposure to emissions from biomass fuels in smoky kitchens.

The broad consequences of the fuelwood shortage are therefore of global significance, and should be incorporated into policy design rather than be seen as solely women's concerns.

CHAPTER 3

Historical trends and their effects on women's links with forestry, an overview

The consequences of deforestation are complex, since environmental repercussions are interwoven with socio-economic problems. Recent economic trends and development policies are primarily responsible for the present situation. But whatever remedial efforts are undertaken, they must take account of the historical developments which set the whole process in motion. This process began centuries ago.

Women's healthy links with the forestry sector existed within a context of pre-colonial survival systems. Indigenous production arrangements prior to the introduction of plantation agriculture were built on a strong sense of continuing interdependence with the local environment. This manipulation of resources for subsistence was a responsibility shared by both men and women and it implied a concept of local self-reliance. Family subsistence was ensured through agricultural activities in paddy lands, rain-fed dryland plots ('chena'), and homegardens, as well as through the use of forests, common lands and reserve areas.

There is a paucity of historical data pertaining to the pattern of labour division between men and women in these crop production systems. However, rural households display a marked gender disparity in time-use pattern (see Figure 5). Rice cultivation in rain-fed or irrigated fields historically was done primarily by men, while women were heavily engaged in dryland farming. From the data available, however, we may conclude that women contributed 32% more labour to family units than did men (Wickramsinghe 1993f).

Figure 5 *Gender Disparity in the Seasonal and Annual Time-use Patterns*

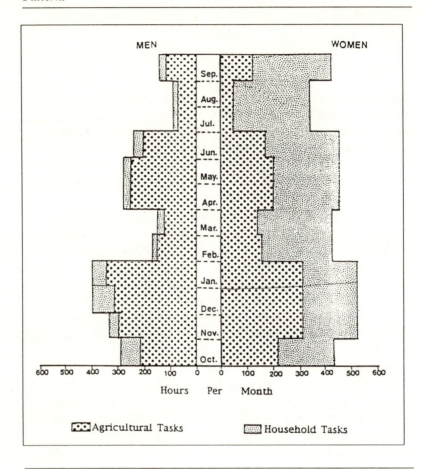

Source A. Wickramasinghe (1993f)

The activity-based division of female labour as reproductive and productive is rather vague, firstly, because crop production for the family is part of reproduction activities and, secondly, because women often attend to both productive and reproductive activities simultaneously. Therefore no rigid division of tasks can be expected. Home-focused and home-based work was a condition of smooth household management. What is interesting is what happens to women's control over resources when they fall into the domain of private ownership, which is a key factor in the deforestation process.

Expansion of commercial agriculture

In Sri Lanka, large-scale clearance of forest lands began more than a century ago with the beginnings of the plantation system. This created not only environmental changes, but ruptured the economic systems of local communities and the survival systems of people.

First, it wiped out the pre-colonial subsistence economy. Until the British arrived, rice was the mainstay, along with some other crops like coracana, root crops, vegetables and tubers. The valleys were cultivated with paddy to produce the staple diet, while the uplands supported homegardens. Hillsides were skilfully terraced, while forested areas were well looked after to provide protection, security and survival needs including water, woodfuel, fodder, medicinal products, food, fibre, home-construction materials and other products. There was a communal interest in protecting the commonly-owned lands around villages, as their benefits were also shared (Pieris, 1956).

With the arrival of the Portuguese in Colombo in 1505, commercial crop-production was recognised as a great source of cash revenue, but the Portuguese cinnamon trade did not damage subsistence agriculture. Later, the Dutch were able to organise the cultivation of cinnamon and coconut for export. However, it was only after the fall of Kandy in 1815, when the British took over control of the whole island, that steps were taken to radically alter the economic base of the country through the expansion of the plantation system in the wet areas (Perera, 1984). Well-distributed rainfall and diversity in altitude and temperature provided ideal environmental conditions for plantation crops. In addition, the highly productive soils, which were neither utilised nor exposed, and were enriched by the forests, proved advantageous for the establishment of new crops.

In the 19TH century, the country's entire economy became commercialised and export-oriented. In the dry-zone areas, where shifting agriculture had been predominant, subsistence production was replaced with market-oriented crops. About 9,000 hectares of land in the mid-hill country were cleared of forest to make way for coffee plantations. This spread to about 125,000 hectares until, in 1886, a leaf disease wiped out this industry. By the beginning of the 20th century, coconut covered 393,000 hectares of former forest land in the low parts of the country. By 1935 this had risen to about 445,000 hectares. The most dramatic deforestation took place in the wet zone highlands, where tea and rubber plantations took off at the end of the nineteenth century. Tea cultivation expanded from the Kandy plateau to much higher elevations covering the Nuwara Eliya and Badulla districts, while rubber expanded in the west and the south, where rainfall and temperatures are high. These crops are still the major staples of the country's economy; in 1980, tea, rubber and coconut provided over one-half of its export earnings.

This pattern of commercial agriculture weakened the subsistence sector and destroyed the prevalent system of communal land ownership. Villages shrank and the labour market expanded, forcing indigenous people to join the plantation labour force or suffer from the inadequacy of resources to support their communities. Women, whose production activities were focused on family and home, experienced greater marginalisation, due to the difficulty of adapting to the new cash economy while sustaining their households by purchased commodities. In fact, as has been shown by Carla Risseeuw's (1988) study of pre-colonial gender relations, the position of Sinhalese women prior to European colonial expansion was preferable in many respects.

This pattern of commercial agriculture permanently affected rural survival systems so much that after four decades of Independence, a substantial proportion of the basic food commodities are still imported via export earnings of tea, rubber and coconut. In addition, with the increase in population (17 million inhabitants in 1990), land per capita fell to 0.38 hectares and resources have become steadily degraded. As people in rural areas go in search of wage labour, their ability to supply their own basic needs has been diminished, so they are forced to overexploit the resources which are left. The greater socio-economic instability which accompanied these trends has not been addressed.

This economic pressure has had major effects on land-use practices, which have changed as a result of mass production and commercial agriculture. In most areas this did not suit the topography, underlying geology, soil and rainfall. Thus, the environmental degradation which commenced with deforestation has advanced at an alarming rate, endangering the future of the economy as a whole and rural survival at large. Unable to ensure household subsistence, rural women have been forced to rely on free food subsidies and the irregular cash earnings of men.

The dismantling of the forest cover

Increasing pressure on land for agriculture as well as for settlements and other activities has accelerated deforestation. By 1981, the country's population reached 15 million, and land density was 227 persons per square kilometre. This implies an increasing demand for land and over-exploitation of many resources, including forest lands. According to the information available for the year 1982 (Seneviratna, 1982), the present forest resources are estimated to total only about 1.62 million hectares, of which 1.3 million lie in the dry zone (Figure 6). In the densely-populated wet zone, where tropical evergreen forests flourished, only 0.593 million hectares remain. Thus, the forest cover is less than 25% of the total area of 6.6 million hectares. Forest inventories show a diminution of over one-half of the forest cover of 2.9 million hectares since 1956.

Reafforestation has been taking place, but on the basis of different priorities, and at a slower pace. Forests are constantly being lost, while capital is being invested in reafforestation. For instance, about 300,000 hectares of dry-zone forests were recently cleared for irrigated paddy under the Mahaweli Development Project while reafforestation was being undertaken elsewhere. Then, there has been the conversion of low-productive dry-zone forests into productive forest plantations. Some forests have been completely cleared to establish teak monocultures (*Tectona Grandis*) in Monaragala, Ratnapura and Matale. Similar types of deforestation are also related to some community forestry projects. Forest areas have been cleared for this purpose in Monaragala and Badulla. Although these transitions are scattered and small-scale, they are responsible for the disappearance of forest resources, forest products and the wealth of biological resources.

Figure 6 *Distribution of Forest Cover in Sri Lanka in 1981*

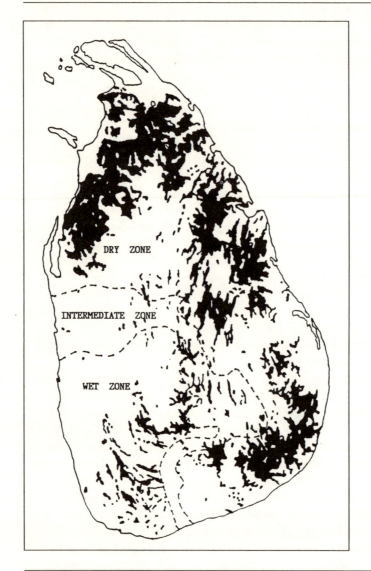

In addition, in some scattered locations, private owners have been forced by legislation, for example the Land Act of 1979, to clear natural forests. One such example is the clearance of eight acres of natural forest in Digannewa village, owned by 42 members of one family for four generations. This acreage was a patch of healthy vegetation for centuries and a source of medicinal herbs and indigenous species in the area. Although the owners wanted to protect it as a family resource, they were powerless to act against the law. In December 1989, they had to clear the whole plot to prevent the state acquiring the land for cultivation. In these cases, the law is implemented in a narrow way, and such incidents do not draw the attention of policy-makers to the need for conserving forests with community participation.

In other cases, lack of political will to implement the law is the reason for forest denudation and degradation. For example, the cultivation of cardamom in the Knuckles Range Forest sacrifices ecological considerations to the economic benefit of earning a substantial amount of foreign exchange on cardamom export.

Women's marginalisation

In spite of the revenues collected on the export of these crops, household subsistence, particularly of the resource-poor, has suffered from insufficiency of cash returns. With the demarcation of property boundaries, women's authority over common lands has diminished. In some plantations in the wet zone areas, women have become providers of cheap labour. The national statistics focus only on this waged contribution by women to agriculture, ignoring the important work they do in the informal subsistence sector. As shown in Figure 7, if national statistics are followed, then one has to conclude that nearly 68% of the women in agriculture are in plantations. The bias towards the formal sector gives the impression that, with the denuding of vast areas of forest lands, women's labour has been absorbed into the plantation sector and overlooks their contribution to the subsistence sectoral economy in which more than 70% of rural women are involved.

Figure 7 *Distribution of male & female labour in agricultural &
fishing sector (%), 1981*

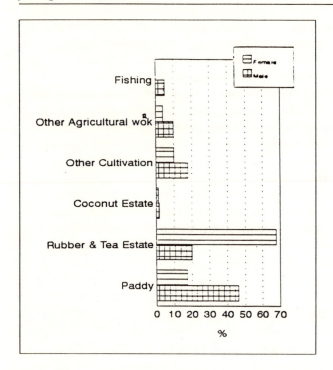

Source Wickramasinghe, A. (1988a)

If we compare the data in Figures 8 and 9, we see that women in
subsistence sectors are not considered to be engaged in production
work. This is largely due to the bias in recognising women in the cash
economy and to the conventional thinking that women's
contribution to family subsistence is to be taken for granted as a part
of their domesticity.

Figure 8 *Spatial Difference in the Female Activity Rates in Sri Lanka in 1981 – By districts*

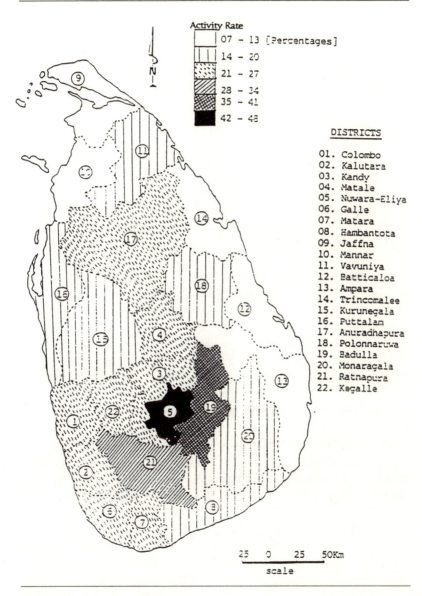

Activity Rate

	07 – 13 [Percentages]
	14 – 20
	21 – 27
	28 – 34
	35 – 41
	42 – 48

DISTRICTS

01. Colombo
02. Kalutara
03. Kandy
04. Matale
05. Nuwara-Eliya
06. Galle
07. Matara
08. Hambantota
09. Jaffna
10. Mannar
11. Vavuniya
12. Batticaloa
13. Ampara
14. Trincomalee
15. Kurunegala
16. Puttalam
17. Anuradhapura
18. Polonnaruwa
19. Badulla
20. Monaragala
21. Ratnapura
22. Kegalle

25 0 25 50Km
scale

Source A. Wickramasinghe, 1992b

45

The present situation represents, not an imbalance in spatial patterns, but rather an imbalance in survival needs and supply sources. Although the dry zone areas of the country have more forests in terms of area, districts where women have strong links with the forests are in the wet zone areas because the peripheral areas of these forests are characterised by the presence of indigenous communities. This means that attempts at rebuilding forest-survival links need to be initiated in the wet zone areas. No substitutes comparable to forests, or crops comparable to multi-purpose trees, have been introduced into the rural landscape.

Overexposure of land due to seasonal crop monoculture causes degradation of soil, water and vegetation, as well as changes in atmospheric conditions. While the environmental repercussions are being noted, the increase in population has aggravated the pressure on diminishing resources. Poverty is one of the causes for poor resource management, which, in turn, accelerates land degradation which intensifies the poverty under which women are struggling. In the absence of desirable strategies, economic policies and political will, the vicious cycle seems to continue. Women in rural communities are at the centre of this cycle, due to their lack of control over the decision-making process.

As has been pointed out by Blaikie and Brookfield (1987), all aspects of the relationship between land degradation and society are both social and physical. Both socio-economic and environmental implications of deforestation are well-marked in areas where rural livelihoods are heavily dependent on land resources. Problems are acute in areas where the land is most heavily utilised. The forest areas which were not exploited in the plantation system have been subsequently eroded by commodity production with the penetration of the market economy. Private capital, with its subsidies and inputs, has been able to attract subsistence farmers, their land and labour.

Figure 9 *Distribution of Cultivated Land by Subsistence and Plantation Sectors in Sri Lanka*

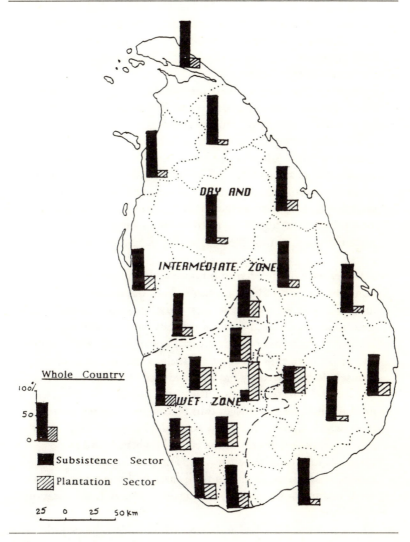

Source A. Wickramasinghe, 1992b

Development policy has had the following results:

1. Segregation of conservation from other sectors. Forest conservation policies, backed by the law, have excluded other sectors like agriculture and contributed to the problem. These have eliminated the people's engagement in conservation and their customary rights and access to the forest, without examining the nature of their links.

2. Emergence of a rather compartmentalised forestry sector, which places the responsibility of protecting forests and reafforestation on a technically-trained group of foresters. This has displaced people who were genuine forestry experts by virtue of their cultural systems. Adoption of costly and externally-improved tree species, and strategies unsuited to local conditions, which results in the transformation of community concerns and loss of confidence. This compartmentalisation breeds the feeling among local people that 'forests are the concern of the state administration, not for the people; and people are not for the forest';

3. Changes have occurred within households and communities. The non-availability of diverse forest products has reduced the multiplicity of the household diet. Nuts, fruits and honey, which were collected and sold on local markets have also diminished. Medicines, fibre and other industrial raw materials, like bamboo and rattan have either become difficult or impossible to obtain. Household food security has been reduced and food habits, income and forest based-industries have altered. Sources of many products have been reduced, thus placing greater pressure on those which are left such as the Knuckles range of forests and the Adam's Peak Wilderness in the central highlands and the Sinharaja forest in the South-west. The inability of recently-established forests to fulfil these multiple requirements tends to antagonise rural women against modern reafforestation programmes.

4. The scarcity of woodfuel and water which affects the majority of people directly or indirectly, and women's time and energy are overstressed as a result.

5. Deterioration of women's and families' health and the nutrition status.

6. Acceleration of soil degradation through seasonal cropping.

48

7 Monoculture production has created greater dependence of farmers on the market and made them more susceptible to market fluctuations. Self-supporting life-styles are vanishing.

8 Income differentials are growing, as the wealthy, who can invest in mass production, are able to participate in the large-scale commercial economy, while small-scale farmers have become marginalised. Landlessness among the poor is growing, as common lands become privatised.

9 All these transformations have reached a critical stage. The forest cover of the country has been reduced to about 25 percent of the land area. At present most of the remaining forest cover is in the same dry zone areas. This shows a greater spatial imbalance in the distribution of sources of supply and demand.

Women's struggle against resource-depletion

Researchers have offered different interpretations of women's resistance to forest destruction and have made diverse assertions about the nature of ecological feminism (Karen Warren 1987, Van den Hombergh, 1993). In commenting on women's actions in India, Vandana Shiva (1989) emphasises the links between women and nature. This might be interpreted as a biologically-based link, which obscures the main reasons why women took part in the protest against logging. It is more realistic to observe that women protest against the clearing of forests for timber and land because of their knowledge, needs, problems and concerns. It is a material interest in the well-being of their families which motivates them, not a vague biological urge. Despite increased effort, women in rural areas simply cannot find substitutes for the resources they are losing through forest destruction. Their capacity for adjustment is being taxed to the limit. The costly alternatives offered through development programmes, like electricity and pipe-born water, remain unattainable luxuries.

CHAPTER 4

The loss of women's control over rural resources

The findings of field investigations conducted in two villages, *Bambarabedda* and *Madugalla,* in the intermediate zone of the central highlands reflect the situation in hundreds of villages in that region.

The village setting

Although the mainstay of the economy of these villages is subsistence farming, the location of well-established tea plantations in the upper part of the territory restricts the areas available to subsistence farmers and limits their access to, and control over, resources. As paddy cultivation is confined to less than 20% of the land, the possibilities of producing rice, the staple food, is extremely low. Cultivation of tobacco and vegetables on the steeply sloping terrain is extremely important as a source of employment and income. While subsistence farming produces food for family consumption, seasonal casual employment is the primary source of income of resource-poor families. This employment affects the share women have in the control of resources. Men migrate away on casual work while women produce crops and manage homegardens and households. The gap between rich and poor is wide.

Overexploitation of resources results in reduced soil productivity, and increases dependence on artificial inputs. As a result, men's control over commercial agriculture has been strengthened, since they have greater access to artificial inputs.

Villagers have been influenced by the investment in commercial crops by private capital. Traditional subsistence living has deteriorated, almost to the point of collapse. Forest lands once shared by the villagers are now under private ownership. Increasing land degradation has reduced interest in investing in them and, at present, some areas have been abandoned. Poverty has been aggravated by the non-availability of employment and the inability of the subsistence sector to re-establish itself in the deteriorated environments.

The greater susceptibility of the land to degradation is evident in the physio-ecological conditions. The occurrence of rainfall within a specific period in which farmers turn the soil or plough the steeply-sloping land increases the susceptibility to erosion. Without vegetal cover, the process has a devastating effect. Cultivation is confined to the period of rainfall because of the land's low moisture content and the drying up of springs and the water sources shortly after the termination of rainfall. Productivity is reduced, due to the use of undesirable land-use practices and strategies unsuited to the sloping terrain. The forests that once covered the upper slopes above the clusters of houses are now exposed.

Because of the non-availability of substitutes for forest products, seasonal scarcities in food, fodder and fuelwood are acute. The land degradation has affected the distribution of perennial vegetation. Most of barren slopes are now free of perennial crops, and the women are heavily dependent on riverine vegetation, hedges and isolated forests for fuelwood, fodder and some food products.

Socio-economic organisation

The two communities in these villages have a history which goes back a century before the plantation sector was established. Their low exposure to the outside world explains some cultural and social norms regarding the division of labour and villagers' social links. Average family size is quite large – 5.4 persons in Bambarabedda and 6 in Madugalla (Wickramasinghe, 1992c). Hence, at least 10.8 kg of fuelwood is needed per family per day in Bamabarabedda and 12 kg in Madugalla. If women take the sole responsibility of gathering fuelwood for domestic cooking, then they have to gather wood 3 to 4 times weekly, allowing a maximum of 21 to 24 kilograms per headload. The health consequences related to carrying excessive weights of wood has not been taken seriously (Wickramasinghe, 1993g and h).

It is impossible for women of the small farm families to meet family needs from their household lands. The women of better-off families, whose family lands are about 10 hectares or so, spend less time and energy in collecting fuelwood because they hardly need to carry headloads. Most households with less than 0.5 hectares of family lands suffer from a double burden. The first is seeking livelihoods outside their farms and the second is dependence on external resources for food, fuelwood and other materials. These burdens have fallen heavily on the shoulders of women and the source of household income and wealth affect women's engagement in extra-household tasks. For instance, carrying fuel-wood on headloads is never performed by women in the better-off families, as they can hire labour for such cumbersome tasks.

Household structure and survival tasks

In these traditional villages, the family is the unit of survival, operation and production. Family structures have a great influence on women's activities. Three types of structure are evident. Nearly 72% of all households are nuclear families. The single-parent women-headed families amount to 11%. The remaining 17% constitute extended families. Within a family men and women have complementary tasks. The tasks performed by women are primarily geared towards household maintenance (see Table 5).

Women's engagement in family survival activities is rigid because of the cultural ideology which dominates their lives and their lack of mobility. Women are involved in subsistence agriculture, caring for the household and children, household crop production, animal husbandry, and provision of tree products, processing and utilization.

Male roles, traditionally related to the cultivation of paddy and dryland farming on the hillslopes, have changed because of the expansion of tea plantations and commercialisation of agriculture with the introduction of tobacco in the 1960's. There has been an increase in women's responsibilities over home maintenance and household crop production. At present, males either concentrate heavily on cash crops if the family owns an adequate area, or work on a wage basis for the landowners or plantations in the neighbourhood.

Table 5 *Gender Specific Division of Activities Related to Household Maintenance*

Activity	Female	Male
Food processing and preparation	88	12
Fetching water	92	8
Collection of firewood	69	31
Collection of fodder	50	50
Preservation of tree food	94	6
Keeping house and yard	92	8
Producing vegetables, tubers & fruits	84	16
Raising children	80	20
Bathing children	75	25
Attending to sick in the family	80	20
Seeing to children's' education	68	32
Washing clothes	98	2
Caring of elderly	90	10
Rearing animals	60	40

Source field information (1989).

In the labour market women are less important, for several reasons. The first is the bias in the family structure which assumes that men have to support their families as the head of the survival unit. The second is the ideology of women's domesticity, which is an obstacle to their involvement in outside work. Finally, women are unable to work outside their homes due to the non-availability of substitutes for the care of children, food preparation, fetching of water and other domestic work. This situation exists with regard to nuclear families. The domestic work falls less heavily on women in extended families, since other members lend hands to do housework, including gathering fuelwood and producing crops. However, women's participation in paid work is high in small families with grown-up children. As was noted in about 12 cases in Bambarabedda, women in large families with small children keep their daughters at home instead of sending them to school so that they themselves can go on paid work. Women's work in the domestic domain includes a number of tasks related to forest/tree uses: gathering of fuelwood, fodder and processing and preparing of food products.

Restricted employment opportunities and strong seasonality in agriculture have made women concentrate heavily on activities based at and around their houses. As outside paid work and household farm activities fall in the same season, male partners work outside their homes more often, while women, with the help of children, cultivate their own land in between the other domestic work. For women, farm work includes preparing fields, raising nurseries and planting harvesting products, nurturing of trees, gathering wood and crop residues and growing crops. In about 48% of the cases, women's farm work is fragmented as a result of having to attend to children, housework and cattle.

Women exercise power in decisions on family budget, domestic crop production, tree farming, food consumption and children's work etc. Under the increasing economic pressures, the husband's wage is not adequate to cover the cost of food. This explains women's priority for forest and tree-borne food to supplement family needs. In about 68% of the total households surveyed, forest foods and subsistence food production are the main contributors to the family pot. This enables women to spend earnings on fulfilling other family needs. The reductions in family food expenditure by utilising forest substitutes has been well recognised as a mark of a women's efficiency in managing households. There is a rigid division of household tasks, in which women do domestic, farm and forest duties, but with the advancement of cash crops, they also try to earn cash to support their families. A similar situation has been described by Schrijvers (1985) in her study in the dry zone.

There are no structural adjustments in the division of household tasks to accommodate the changes in environment. Women work longer hours to assure family well-being since the handling of forest resources are not considered a male concern. About 28 of the women surveyed said men are not competent at domestic tasks and are unable to undertake multiple activities including the processing of tree products. For women, the best substitutes for rice are fruits of jack (*Artocarpus heterophyllus*) and bread fruit (*Artocarpus altilis*). If they do not use the tree food, not only will the products go waste, but family food consumption will also be reduced. According to 72% of the women in the survey, their flexibility and efficiency at combining tasks enable them to do more work within the course of a day. This acceptance of simultaneous multiple tasks explains why they do not have specific time allocated to nurturing trees and processing products. In most cases, it is not possible for this unit of operation to function without women's contribution and control. Women are the gatherers of tree products. If they are not able to use non-wood products of forests and trees, then such outputs get wasted, while trees of the farms and homegardens will become stocks of timber.

Expansion of commercial agriculture into the village economy

The recession of the wet evergreen forests of the nearby Knuckles range has been a great loss to the people. The process started with the clearance of forests for tea cultivation at the edge of the villages which became reduced to clusters of houses separated from other areas, particularly the forests. Their survival became isolated from the forests and restricted to a limited resource-base. The impact of these transitions on the lives of women is more significant than that of the men, because their engagement in managing forest resources for subsistence has been marginalized.

With the acquisition of land for tea plantations, village labour gradually entered the plantation. Increasing population pressure and scarcity of resources forced villagers to seek outside employment. There was no land available for the expansion of dryland farming, which is done on a shifting basis. The area available for crop production became limited to a few stretches of paddy along the riverine areas, homegardens and a few plots of farmland. With the fragmentation of land, people who had a small plot became landless. Bambarabedda was severely affected because all the hill slopes above the village were taken for tea plantations.

The situation in Madugalla, which is located on the left bank of Ma Oya, was slightly better because the area which did not come under tea plantations was much bigger, due to topographical factors. As a result, the villagers had the advantage, keeping about 497 hectares, comprising 34% of the land under paddy, which supplied their staple food. In Bambarabedda, paddy cultivation was limited to 33 hectares and accounted for about 9%, causing widespread poverty. The unequal distribution of land among families is attributed to class structure. Families cultivating a comparatively larger amount of land with paddy, a homegarden and farm plots were better off than the ones who had a homestead alone.

Families with very limited farmlands were no longer able to meet their increasing survival needs by subsistence farming. Food scarcity and family hardship made male members seek some paid work. In the early 1950s, as related by the elderly of these villages, about 30 males in Bambarabedda and 12 in Madugalla sought casual employment in the plantation sector. Migration of labour from the subsistence sector was the primary reason for changes in the rural survival systems, the marginalisation of women's experience and the multiplication of their responsibilities. The pattern of mutual sharing turned into one with strong divisions, encouraging men to migrate in search of cash work and women to work free of payment in farming.

The absorption of male labour, particularly of the poor, first resulted in strengthening male autonomy over cash earning. This was an important transition in the rural survival system, because people had to depend more and more on cash when land fragmentation reduced their holdings. At the same time, women's dependency on men's earnings increased. However, in the late 1950's, the increasing labour required for plucking tea and weeding plantations absorbed women of poor families into the plantations as well. This helped reduce their dependency on male earnings, but brought about conflicts in the plantation sector when women were made to work for low wages under the supervision of men. This organisation of waged work was a landmark in establishing men's control over women's labour, with the adverse consequence of removing women's autonomy in exercising their managerial capacities developed over generations of farming. Under socio-economic pressure, women had to find work along with their male partners to support their families. The physical exhaustion of the long working day, together with the duties of family crop production and the establishment of boundaries to formerly-free resources, no doubt proved extremely stressful.

Land degradation, which weakened tea production, did not remove the influence of the cash economy on village activities. The penetration of tobacco cultivation into these areas in the 1960s was for the purpose of revenue collection and not for rehabilitation. Small-scale farmers who cultivated a substantial area of land with food crops were attracted by the extension services, subsidies and inputs offered by the Ceylon Tobacco Company. At this point, the impact of tobacco was greater in Madugalla, because the land area available for this crop was bigger. Increasing land degradation on the steeply- sloping terrain over decades reduced the casual work offered in tea plantations to Bambarabedda villagers and, as a result, much of the labour was displaced. Only 12 tobacco barns were set up here whereas, in Madugalla, the number was over 150. However, possibilities of earning cash and the efficiency of services delivered at the farm level made even the families with small farm plots adopt tobacco as the main crop, displacing food crops. They became the suppliers for the large barns and received subsidies through them. In terms of labour input, most of the production tasks were covered by family labour. There was a demand for women's unwaged family labour. Unlike in the traditional systems of cultivation, where reciprocal exchange of women's labour is predominant in transplanting and harvesting, no such practice existed in tobacco. Women who once made decisions on resource management now provide barn labour, and engage in raising nurseries, plucking and grading tobacco leaves.

Homegardens, however, remained a place of autonomy for women. Tasks considered strictly male are limited to climbing of trees to pluck fruits. With the reduction of forests, women's focus on homegardens has increased and they are geared to fulfil the daily needs of the family, although their capacity is determined by the extent of land available. Mostly, homegardens today accommodate many types of multi-purpose trees, tubers, vegetables and also support farm animals. Homegardens are undoubtedly the most attractive unit of production for women.

A number of other changes followed. Under the twin pressure of commercialisation and the reduction in subsistence crop production, there was an increase in female agriculture wage labour. In Bambarabedda, except for 12 families, women in all households work for a daily wage if work is available. A man's daily wage – between Rs.75-100 in 1990 – is not adequate to buy the family food requirements. Most women work on the bigger peasant farms, especially in raising nurseries, transplanting, weeding and harvesting of tobacco and paddy. They are used as a cheap source of labour. About 40% of them are engaged in tobacco cultivation either as family labour or as wage earners. Yet they are practically cut-off from decision-making. The first reason is that they are not considered the main target group of the Ceylon Tobacco Company. This is mainly because cultivation is done on instructions of the field supervisors given to the land owners, which means that women do not have any direct links with the company. The second reason is the conditions laid down by the company regarding eligibility for subsidies and credit. As the ownership of farmlands is in the male's name, the company treats males as the focal point, not only in distributing materials like seeds and fertiliser, but also in signing contracts, and delivering training and instructions. As a result, men exercise control over women's labour, both hired and family labour, since all cultivation has to be done according to the instructions issued by the company. Unless farmers follow these instructions, they forgo the chance of getting assistance or inputs. These circumstances have led to the strengthening of male supremacy in the interest of commercial agriculture.

This state of affairs has strengthened men's engagement in trade too. Here again, transporting of raw or dried tobacco is done by men, due to their direct business contacts with the company. As the males have legal titles to the land, it is they who sign documents pertaining to supplies of produce. Women's backwardness in education, communication and in handling trade as a whole also contributes to this situation.

Insecure village survival

Women recognise the insecurity of this type of monoculture farming. They know that their lives have changed from what it was when they lived in their own habitats. They remember the less cumbersome lives of their mothers (see Table 6). They still have the knowledge transmitted by their grandmothers and are aware of how important the natural environment is to their lives. Unless the conditions of these habitats are improved and resource depletion is halted, they know, that living in these terrains will be a struggle.

'This whole place is like a desert,' one village woman observed. 'We have no water during the dry seasons. The water level of the river also goes down. As it is used for bathing and washing, the water is polluted. Then, to get fuelwood to boil water we walk about 1.5

kilometres and spend about 3-4 hours a day. Most of the food commodities we get from outside. To supplement family income, we work outside, but it is not easy to find work.'

According to their memories, almost all the trees which flourished in the farmlands were felled in adopting tobacco. Most trees mentioned by women are multipurpose ones and include margosa (*Azadirachta indica*), mango (*Mangifera indica*), mee (*Madhuca longifolia*), jack (*Artocarpus heterophyllus*), bread-fruit (*Artocarpus altilis*), guava (*Psidium guajava*), arecanut (*Areca* catechu) and other bushy species. This depletetion has emptied the smoke trays in kitchens. Along with this change, the oil extracted from the kernels of Mee (*Madhuca longifolia*) was not available for cooking or medicinal purposes. Most of the food processed during the fruit producing seasons has disappeared, and now much has to be purchased at a higher cost.

Rice consumption has increased due to the fall in tree food availability and root crop production. For the poor, it is difficult to afford even a coconut per day. This tremendous transition has resulted in a need to spend more and more energy and time to find raw stuff for a family meal. Instead of reducing women's workload, commercialisation has increased the pressure and created dependence on food from the market and the necessity to earn money for this by working for wages 20-30% lower than those of men. To handle this increasing number of tasks women must sacrifice leisure and sleep, reducing their sleeping time in peak agricultural seasons to five or six hours.

The high consumption of fuelwood in curing tobacco has increased land degradation and the pressure on resources, these women note. Competition for fuelwood between domestic and industrial uses has turned into a competition between women's needs and those of the cash economy. About 4,000 kg of fuelwood is used for curing 1,000 kg of tobacco, (Wickramasinghe, 1990c) and in 1990, the price was about Rs.475.00. This price represented a threefold increase in about five years. In earlier decades, local resources had satisfied 100% of fuelwood requirements. Between 1970-80, this dropped to about 60-70%. Now, fuelwood is provided by the Ceylon Tobacco Company. The locally available hardwoods of homegardens and hedges are utilized by the tobacco barn owners because these are cheaper than the wood supplied by the company. Men are responsible for getting industrial fuelwood, while domestic fuelwood gathering is left to women who have to resort to the illegal collection of twigs from the fallow bush of state lands. This practice creates severe domestic hardships.

In addition to the time and energy spent in gathering fuelwood and fetching water, 4-6 hours are spent on cooking meals per day. Although it is very difficult to quantify the resulting adjustments that women have made, in a sample of 24 families the number of pots cooked for family consumption has been reduced by about 30%. About 40-50% of the families do not use boiled water due to the problems of getting fuelwood, although they are aware of the advantages of drinking boiled water. About 69% who used parboiled rice before husking have abandoned the practice completely because parboiling requires large amounts of better quality fuelwood (Wickramasinghe, 1991c). Parboiling of rice is preferred to raw rice because it increases the quantity. Smoke drying of jack, bread-fruit and cassava have been completely given up.

Lack of raw materials to make household utensils compel the women to buy sleeping mats, winnowing fans and drying mats from the market. With the scarcity of raw materials, women's technological knowledge and skills also disappear, resulting in a greater scarcity in household utensils while increasing the need for cash. The inherited skills passed on from mother to daughter are also diminishing because species like rattan, bamboo and pandanas are extremely scarce.

To cope with these changes, women have increased the multiplicity in their homegardens tremendously, integrating more and more trees, herbs and crops, including coffee, pepper, banana and vegetables. With the displacement of trees from their farmyards, women have tried to grow many varieties using the available land space. According to them, it is the most productive system, offering a multiplicity of products that require neither costly inputs nor much labour. Women report that most of the crops are compatible and survive longer. Residues of the leaf biomass and regeneration processes keep the system sustainable. Organic manure like cow-dung and *Gliricidia* and keppitiya (*Croton laccifer*) are adequate to give a good harvest of vegetables and tubers. Obviously, they are basing their practices on a sound knowledge of ecology and with a view to protecting the environment and pursuing sustainable agriculture.

Table 6 *Women's perceptions regarding sustainability of family production systems in Bambarabedda and Madugalla.*

System	10 years ago	Current	Future 10 years after
Paddy	similar to current situation (100)	cost has increased (80) similar (20)	cost will increase (100)
Home garden	similar to current situation (70) less productive(30)	multiplicity and utility will increase (80) no change (20)	will be most productive (80) reduce productivity(20)
System based on tobacco	very profitable (100)	continuing with a marginal profit (70) continuing without a profit (30)	will collapse (80) continue without profits (20)

Source author's field work (1990).

Table 7 *Predominant perennial species in the homegardens in Bambarabedda and Madugalla*

Species		Uses				
Local name	Latin name	food	medicinal	fodder	fuelwood	timber
Kos	Artocarpus heteraphyllus	*	*	*	*	*
Pol	Cocos nucifera	*	*		*	*
Del	Artocarpus altilis	*			*	
Mi	Maduca longifolia	*	*		*	*
Siyambala	Tamarindus indica	*	*		*	*
Alipera	Persea gratissima	*			*	*
Amba	Mangifera indica	*			*	*
Puwak	Areca Catechu	*	*		*	
Papaya	Garcia Papaya	*				
Kohomba	Azadirachta indica		*		*	*
Pera	Psidium Guajava	*			*	
Karapincha	Murraga Koenigii	*	*			

Species		Uses				
Local name	Latin name	food	medicinal	fodder	fuelwood	timber
Glircidia	Gliricidia Sepium			*	*	
Pepper	Piper nigrum	*	*			
Cocoa	Theobroma cacao	*			*	
Coffee	Coffea spp	*			*	
Jambola	Eugenia jambos	*	*			
Anoda	Annona muricata	*			*	
Lunumidella	Melia dubia				*	*
Delum	Punica Granatum	*	*			
Kitul	Canyota urens	*	*		*	*
Kesel	Musa sp.	*				
Rambutan	Nephelium Lappaceum	*			*	*

Source author's field inventory (1989)

Women's priority – homegardens

Homegardens in these villages look like patches of forest on barren slopes. These are places where women can exercise autonomy. They demonstrate the concepts they hold regarding agroforestry and the liveable environment. Shade, coolness and psychological relaxation are the benefits that are enjoyed. The reasons for accommodating perennial species are related to these as well as to the products. Species planted are carefully selected, particularly with an eye to the multiplicity of products to be gained. Diversity in homegardens has been increased through domestication and nurturing of wild species. In these two villages, as in many rural communities in Sri Lanka, there are no households without a homegarden. Even the near-landless maintain a few trees around their dwellings. As a symbol of establishment and security of the survival systems connected with them. There are more than 62 perennial species in these two villages. Out of these, about 60% are introduced while the remainder have entered through natural processes, primarily seed carriers and wind.

Even within the villages, variations in homegarden composition are outstanding. Women plant more and more species closer to homes, so that the intensity of species declines the further you get from dwellings. Vertically, the diversity of homegarden is structured by light penetration. It is managed as a system of food procurement (Wickramasinghe,1991a). To a great extent, the multiple products of the homegardens are similar to that of forests, that is, they include food, fuelwood, fodder and medicinal products. Most of the predominant species are of multiple products (see Table 7). Primary combinations include food, fuelwood and medicinal products. The driving forces behind women's greater engagement in home-gardening are the products and the creation of healthy environments. By integrating trees, crops, animals and herbs into a single unit, women intensify the system. For women, the greater the biodiversity, the greater the stability. Species like coffee and pepper enable them to produce some for market as well.

Out of the three production systems – homegardens, paddy farming and dryland cultivation – the women stated that they give priority to the homegarden. In about 80% of the cases surveyed, this was due to their knowledge and complete authority over decision-making in all their activities, and their independence from artificial inputs to keep the system under their control. From the experience of about 20% of the women, they expect that this system will be disturbed with the fragmentation of land among children for the construction of houses in the future. However, this tree-based multiple land-use system is the one in which they work with great self-reliance and confidence. Obviously, their close contact with homegardens is a reflection of their contact with forests.

Whatever the advantages women see in the system of homegardens, the area coming under homegardens is low, amounting to 18% of the total area in Bambarabedda and 11% in Madugalla. Women are unable to expand this due to pressures from the other market-oriented production systems, which are mostly controlled by male owners of the land.

Loss of women's control over resources and its results

Clearly, the situation is now at a critical point. Women are exercising all their capabilities and accepting tremendous stress to keep some remnants of a functional and ecologically-sound system going. But if the situation should deteriorate further this will not be enough.

Many indigenous resource management measures have suffered due to displacement of women's links with the forest. The field evidence brought into focus by women shows that the loss of at least 30-40 centimetres of soil over 5-6 decades has had a tremendous effect on agricultural crop production. Barren slopes with rock outcrops are widespread. Although no experimental studies were conducted in this area to measure the annual loss of soil, the result of a previous study in a tobacco-growing area in the highlands of the intermediate zone in Hanguranketa is of some relevance. In the Maha Oya catchment, for instance, soil loss in tobacco fields was 12-14 tons per annum (Wickramasinghe, 1986, and 1988b and c). This evidence supports field observations and shows the severity of the problems that women have to face if such lands are to be brought under production again. This should be regarded as a serious problem, because many of the peasants have no alternatives to offer the next generation. The potential for growing crops during off-seasons is limited because the soil does not retain water. Soils are too shallow, stony, and poor in terms of organic matter content. In fact, such barren slopes are unable to offer even a bundle of fuelwood.

In addition, almost 60% of the natural springs which were utilized by the local community as spouts for bathing and washing have completely disappeared. During field-research, the elderly were able to locate points where springs and rivers once existed. The seasonality of the feeding tributaries and lowering of the water level of the Ma Oya are remarkable. Women in Bambarabedda now have to reach the bottom of dug wells with coconut shells to fill a pot of water. The situation is not so serious in Madugalla because of available water in narrow valleys between the settlements. At least surface water is available for consumption during scarcities, but women have to walk about 1 to 1.5 kilometres to the valleys to fetch a pot of drinking water. This is an extremely hard task in Bambarabedda, due to the difficult terrain and the narrow foot paths along which water has to be carried.

As the women themselves have realised, they can be more involved in agriculture to increase the diversity and multiplicity which is needed for their survival. Women's priorities include diversification of agricultural production systems, soil and water conservation and getting opportunities to make decisions on forestry and agriculture. They were more interested in becoming contributors to planning,

designing and implementing agricultural strategies than being mere
beneficiaries, since they saw the long-term investment for their
children. But rather than giving priority to environmental restoration,
they now have to work to earn cash and work long hours to meet
daily needs of their families. Deepening poverty and environmental
degradation therefore need to be addressed simultaneously, since they
come together in a vicious cycle.

Recalling agroforestry for sustainability

For women, the success of reafforestation strategies depend on their
acceptability to local people and their suitability to local resources.
The economic viability of existing systems are declining, they say, and
therefore newly-adopted technologies, which increase resource
depletion, are unacceptable. A number of economic pressures have
trapped local dwellers in the current commercial system, via a package
of subsidies, services and market benefits. Private companies have
been able to penetrate into these areas and introduce their crops on
farmers' lands. As the women see it, barren slopes are a result of the
separation of trees from croplands.

The women put forward a number of ideas as to how sustainable
development can be put into practice. The homegardens, they
believe, form the best model of an acceptable system of agro-forestry.
The agricultural crops and trees grown in homegardens together with
its livestock produce a multitude of products. Socio-economic and
environmental requirements are interrelated aspects of this system,
which uses extremely low family inputs and produces multiple
outputs for family survival, includingfood for family, energy for
cooking, fodder for animals, products for other uses and the
development of a liveable, sustainable environment. These ideas
should be examined with a view to seeing how agro-forestry could be
a remedial measure in reclaiming degraded land, improving soil
fertility, infiltration and moisture capacities and satisfying the needs
of the peasants.

Cecelski (1985b) has supported their views in ILO studies,
observing that agroforestry is a promising approach taken by poor
households and especially women, as it replicates the multiple
products of the natural forest and can be practised intensively around

homegardens and more extensively on marginal communal lands. This form of agroforestry is especially relevant to women's primary concern with food cultivation, since production of wood for fuel can be usefully integrated with agricultural systems. Since it is women who suffer the main hardships caused by deforestation, starting with their knowledge is probably the best guarantee of success.

In Sri Lankan villages, the integration of forestry with food crops along with water and soil conservation practices helps local women to re-establish resources. Homegardens consisting of multipurpose species such as jack (*Artocarpus heterophyllus*), bread-fruit (*Artocarpus altilis*), coconut (*Cocos nucifera*), arecanut (*Areca catechu*), mango (*Mangifera indica*), guava (*Psidium guajava*), pomegranate (*Punica granatum*), coffee (*Coffea arabica*), pepper (*Capsicum spp.*) and gliricidia (*Gliricidia sp.*) meet the multiple needs of the community. The lands are not designed, but randomly developed. In most cases, species are freely germinated and nurtured by women.
Short-durational food crops and market-oriented crops, like coffee and pepper, are planted among trees.

According to the information given by the women, the integrity and sustainability of the system are a necessity for its survival. Trees and crops grow in a compatible manner. Rearing animals within the system is appropriate in the management of resources, due to the complementary nature and interaction of the two activities. Homegardens basically include any system of land-use in which woody plants are deliberately combined, in space or over time, on the same land management unit as herbaceous crops and animals (Lundgren, 1987). In general, agroforestry applies to a variety of land-use systems ranging from very intensive farming to extensive pastoral systems (Rocheleau, 1986), and so homegardens covers a range of systems which could be practised in many areas.

A large variety of systems coming under the umbrella of "agroforestry" is not new to rural dwellers, particularly to women. Every plot of land is demarcated with a live fence or a hedge. However, many issues have emerged in promoting it as a desirable land-use system aimed at overcoming social and environmental problems. According to Budd *et al* (1990), the need for integrated planning should be taken into consideration. Agroforestry has its own, specific, vertical role to play, by way of producing a series of important products from vertical stratas not furnished by agriculture.

In addition, according to Farming Systems Research and Development (FSR and D) efforts, agroforestry should also take a horizontal responsibility for maintaining, conserving and, in some cases, restoring areas as a natural resource base. Horizontal benefits are gained on the landscape gradient in relation to the spatial arrangements of the other systems. In other words, farming systems and agroforestry together offer the largest possible range of technical options from which local families can select those which can best be adapted to their own needs and the socio-economic and political context. At present, in Sri Lanka, homegardens represent the most rational system in pursuing these ends.

In introducing agroforestry as a remedial measure, it is necessary to bear in mind that women, as managers of land, carry a number of potential contributions which could be applied in planting, establishing and managing the system. As pointed out by Berry (1977), many of the trees already exist within their land premises because of their usefulness. This is the situation found in this study too. Women's priorities in selecting trees are food, fodder, fuelwood and mulch (Wickramasinghe, 1991d, 1990d). The diversity of the system and the seasonalities in production are well-known to the women.

Much of the empirical evidence (e.g. Saka *et al*, 1990) suggests that agroforestry can provide a sound ecological basis and dependable economic returns. Under the increasing population pressure which has resulted in fragmentation of small-scale farm operational systems into small plots, agroforestry could be taken as an integrated system of production. Agroforestry offers great prospects for restoring the lost resource-base, enriching the soil and stabilising the water resources to rural women in these degraded hills.

However, rather than importing systems developed externally, efforts must be made to enable women to design systems for their lands and maintain them. This highlights the need for drawing on traditional experience and indigenous knowledge to make development efforts acceptable and sustainable to local people in their habitats. It is important to make sure that designs which are developed under specific experimental situations do not supersede the ones which local women prefer in their specific local contexts. Before adopting externally- developed models, it is necessary to recognise the capacity of people, particularly women, to design their own.

Diagnosis and Design (D & D) is the starting point out of which agroforestry technologies are produced in order to fit the needs and conditions of the land users (Raintree, 1987). A lesson learned from investigating farmers' tree-breeding objectives is that their preferences vary according to the ecological niches within which their strategies will be practised (Wickramasinghe, 1992d). The adoptability of the system by the land-users and social acceptability of the system would be determined by the manner in which such systems are able to cater to their survival needs. The Diagnosis and Design approach to developing agroforestry technologies begins with land-users' practical knowledge and objectives. Through diagnosis of the land-use systems, options are identified (Raintree, 1990). Examples where this approach has been applied in Kenya (Scherr, 1990) show the possibility of replicating it as the starting point for agroforestry.

Opportunities and constraints

Close observations of these village situations show that women are not merely a source of labour, but also a part of the society who have a great deal to contribute to the planning, managing and designing of appropriate technologies. In this society, which is culturally homogeneous, women are a vital force in maintaining the family. They contribute significantly to the commercial agricultural sector and farm household lands primarily for the provision of survival needs. Therefore, within the family system, where standards of living need to be maintained, women are the key actors, but they remain in the background because of the acceptance of male headship in the formal arena. In utilising forests for subsistence, although they do not proclaim their contributions, they are the carriers of technologies, including those related to resource management and utilization. They have acquired an intimate, practical knowledge of sustainability, species-usage and ecology in relation to trees and crops. Their experience, knowledge and technology in processing and preserving food is important in maximising use and making the most appropriate choices. Their homegarden management reveals their concern for protection and management of ecological resources, including biodiversity, in the interest of survival.

As Rocheleau (1987) discusses, in many agroforestry systems, women are responsible for planting, tending and gathering, in addition to performing their roles in crop and animal production and consumption. They have been key contributors from the level of identification and design to the levels of implementing, maintaining and evaluation of living projects which are not supported by official intervention. Thus, if development is to be built on local knowledge, the process should be initiated with women. Their ability, knowledge and experience in handling resources like soil, water, trees and other species should be inputs in ensuring the sustainability of systems and adoptability to the society. Women's interest in tree-based production is because of the multiplicity of the accompanying benefits, low-cost inputs and extremely low susceptibility to damage by drought. The incorporation of these users' perspectives is a pre-requisite if people are to be encouraged to contribute the body of knowledge, methods and rationale for technologies. A study conducted in Zambia (Huxly *et al*, 1985) along these lines proved the importance in designing agroforestry on the basis of users' perspectives.

If women are to undertake and play a prominent role, constraints to their participation should be removed. Some of these are long-standing due to the economic and political context in which they have operated. In addition, there are a number of constraints arising from women's own time-allocation patterns. Often they are overburdened with a multitude of household and productive tasks, especially during the period in which agroforestry also needs to be established with appropriate conservational measures. In a situation of poverty, it is practically difficult to invest time and energy in systems which are mostly meant for long-term benefits at the cost of opportunities to earn cash and produce food. On the other hand, diverting women's labour from other productive tasks is problematic, because this could lead to severe problems in household survival. Sharing of household maintenance tasks helps women to be actively engaged.

The most conspicuous problem here is the males' ownership over land, which gives them authority in decision-making. Men can decide to cut trees for cash since they are the legal owners, while women's interest in preserving trees for the benefits of several generations can be ignored. In addition, the establishment of legal ownership over land which was previously freely available robs women further of opportunities to use land effectively and caringly. Then, there is the system of distributing family land among male children and passing over of cultivation rights from father to son. With this strong family-based structure, women do not have a great interest in owing land, although they are the ones who nurture and maintain trees while men take decisions on the harvesting of timber. In general, they are less interested in legal deeds once they belong to the family, the unit of survival. In addition to tenurial matters, social norms, ways of delivering services and assets and in making decisions on trade and land-management make women's status no better than that of labourers.

Women's capacity for decision-making in land management is demonstrated by their methods in homegardens even though they do not own land. When mobilising their knowledge, it is not justifiable to expect them to work for the sake of family well-being, assuming that family approval is a sufficient reward for such an active role. Market-oriented crop production has created some problems, largely due to the procedure followed in delivering extension services from both private and public sectors. The legal owner is taken as the focal point in receiving subsidies and services. Women are displaced in all the formalities followed in the procedures, which are handled by the owner. This situation has become firmly entrenched with the increasing commercialization of production systems subsidised by both state and private sectors. Therefore, male autonomy in market crop production has not only displaced women from their position of decisionmaking, but also wrested control away from them in crop management practices and trade (see also Schrijvers, 1985). As subsidies are received on a signed contract, men can spend the money or use the subsidies without women's consent. Women have no opportunities to obtain credit, even for the purpose of land management, without proving their legal ownership. This limits their motivation for land development or for adopting new technologies. In this context, dual ownership of land use and management seems to be a desirable solution.

Women now have free access to use other land resources, for instance common land, reservations located along river banks and roadside areas, and also isolated patches of forest on mountainous terrain. However, as such areas are located in narrow strips, the possibility of being claimed by the owners of adjoining lands often act as a constraint on the freedom of women. Where such reserves are extensive, with a substantial area of about 0.5 to 1 hectares, women of adjoining lands collect the products including fuelwood. The trees planted to demarcate the farm boundaries of homegardens are harvested by the growers as owners of the 'boundary'. In many situations where valuable timber is produced, the owners of the land on either side claim 50% each.

In the case of Sri Lanka, there is no strong cultural norm which restricts women from planting trees, as has been observed elsewhere, for instance by Chavangi *et al* (1988) in Kenya. This offers opportunities to stimulate women to plant trees. Although individual ownership supported by title deeds is not seen by women as a constraint to planting trees on communal land, they refrain from doing so, due to the anticipated problems of harvesting the products. Thus, they limit their activities to harvesting, which increases the rate of resource depletion. Tree rights on private property are tied up with land ownership, so that the owners' authority is legally required for harvesting timber. Although women themselves do not consider this a problem, their right to harvest timber of the species that they tend is important because such species are investments of their energy and time. The harvest of by-products that they gather is not the ultimate return.

Despite some spatial differences, the literature (i.e. Fortman *et al*, 1988, Chavangi *et al*, 1988, Chinwuda, 1988 and Rocheleau, 1987), which has explored the tenurial position of women with regard to trees, underscores a few important points. First, women's contributions as users and managers of trees are often essential in selecting the best species to fulfil needs and solve problems. Secondly, as they undertake the drudgery of collecting tree products, they have a felt need to grow them within easy reach. Thirdly, as most of the subsistence farming is done by women, they could combine trees into their production systems and tend them while attending to the other crops.

All of these efforts to draw on women's contributions should take care not to destroy the basic structure of the survival unit. As was pointed out by Chavangi *et al* (1988), an intervention approach aimed solely at women would imply that men would be excluded, so that serious conflicts at the household level would probably result. As such situations could surface, programmes should be addressed at the family without segregating women. As is further stressed by Fortman and Rocheleau (1985), it is necessary to be sure that both men and women are involved together so that men's views can also be accommodated. Chavangi's experience of possible conflicts between husband and wife may be case-specific. For instance, Schrijvers' (1985) experience in the northern dry zone of Sri Lanka suggests that, apart from gender, class background plays an important role. Especially in poor households, men reacted positively to projects geared to women. Nevertheless, new systems should not aim to transfer male supremacy to women and displace men and women from their complementary tasks. It is important to solicit men's understanding, support and awareness of the needs as prerequisites of any change. Unless the appropriate approach is taken, women's potential contribution to the whole forestry sector will be excluded.

CHAPTER 5

Forestry and the forestry sector

Forestry policies are focused primarily on forest plantations, conservation and timber production. This focus also includes community forestry programmes which aim, in addition, at guarding natural forests and meeting community needs. The profession referred to as forestry is regarded as a technical one, riddled from top to bottom with male decision-making and operating out of a paradigm in which 'technology' is seen as the key element.

In this chapter, we will assess these policies, not only in terms of practical achievements in reaching their stated goals, but also to see whether modern forestry development efforts have been able to uplift the standard of living of those dependent on forests for their survival. We will also look at what forestry policies have done about equalising the status of women and recognising their roles and potential roles in the interests of development.

Deterioration of the integrity of forestry

The forestry sector was set up more than a century ago with the recognition of a need for protecting the unique biological wealth of the forests, as well as for the purpose of producing timber. As in many other countries, it originally evolved in order to meet industrial and construction needs. An element which was missing, forestry for community needs, was forcefully pushed into it in the 1980s with a number of eye-catching words, such as "Community Forestry", and "Social Forestry".

However, with state involvement in forest management and the establishment of plantations, 'forestry' had become compartmentalised to ensure efficiency in administration and achieve the specific task of timber production. Segregation of agriculture and animal husbandry from forestry in Sri Lanka led to the isolation of trees from their unique multiple function. As a consequence, peoples' links with trees/forests, as well as tree-based survival, was ignored as the state was not interested in preserving these links. While people were expected to fend for their own needs, forestry catered to state timber needs, although fuelwood requirements had been mentioned as a priority.

This focus has resulted not only in a form of segregation within state administration, but also in a clear trend towards delivering services, subsidies and other incentives meant to promote the utilization of land. This segregation is largely responsible for destroying integrated land management practices in which women were heavily engaged. For instance, the services provided by the agriculture sector placed their priority on high-yielding advanced-crop varieties, displacing forestry and marginalising animal husbandry. Similarly, trees came under the sector called 'forestry', whose officials consider neither farmers nor their lands as a part of their domain. As a consequence, to people living in rural areas, the term 'forestry' means state control over 'forests', which, in turn, implies extensive tracts of trees which have been selected, managed and grown in monoculture by this sector. Neither the forestry sector nor the state agriculture sector considered trees to be a means of meeting human needs.

Moreover, in Sri Lanka state policies intentionally compartmentalised crop production systems focused on market-oriented production. Here, the link between trees, crops, livestock and people were overlooked. Integrated land-resource management strategies were confined to homegardens. Systems like this, which integrate species, environmental concerns and people's needs, were ignored until the last decade as 'non-scientific'.

The failure of the sectorally-divided departments to work under one umbrella or have horizontal links resulted in increasing environmental destruction and isolating forestry from other sectors. The segregation of forestry from agriculture led to the displacement of trees from farm yards and vice versa. With the realization of this imbalance, both forestry and agriculture have separately tried to integrate trees and crops in their experimental work. However, they are not formally interwoven or practically linked at decision-making and policy levels.

Later, with the realization of the need for integrating trees and crops to sustain crop production and meet both environmental and social needs simultaneously, efforts were made to restructure both sectors. Since 1976 or so, attempts at integration have taken place under the theme of Integrated Development in which forestry is a priority. The FAO (1983, 1989a and b, 1990a), ILO (1987) and researchers like Hoskins (1980, 1983) and Hoeksema (1989) stressed the advantages of linking these sectors and suggested structural adjustments in the respective sectors. In agriculture, attempts have been made to integrate trees into farmyards (Wiersum, 1990) in order to protect and enrich soil and reduce the cost of production. In addition, the Export Agricultural Crops Department in Sri Lanka has tried to promote species like Pepper (*Piper nigrum*), Cloves (*Syzygium aromaticum*), Nutmeg (*Myristica fragrans*) etc. in farming systems and coconut plantations, with the intention of promoting cash returns and cropping intensity. The animal husbandry sector is also trying to expand this for off-farm income and employment. However, the segregation still exists not only among the state sectors but at the level of community integration.

Common sense forestry

Perceptions of forestry differ widely. Mok (1990) argues, "Forests are for the benefit of the community. They satisfy the basic needs of mankind, maintain environmental equality, conserve biodiversity and contribute to socio-economic development. Unfortunately, there seems to be no central agreement on what a forest is; who or what constitutes a community; and how the forest should be managed for the benefit of the community."

Lack of central agreement has stemmed from lack of common sense. To deal with many of the issues raised here, it is important to ask who manage trees and forests, why they manage them and what opportunities are opened up for people to be continue the human-forest links. Ford-Robertson and Winters (1983) defined forestry to include guardianship, silviculture and human benefits, according to the following principles:

1   Generally, forestry is a profession embracing the science, business and art of creating, conserving and managing forests (forest management) and forest lands (i.e. a forest estate) for the

continuing use of their resources, materials or other forest produce;

2   More specifically (from the producer's point of view), the husbandry of tree crops (silviculture), or (from the economist's) the profitable exploitation of the resources of intrinsic forest land (economic forestry);

3   The science, the art and the practice of managing and using for human benefit the natural resources that occur on and in association with forest lands.

The problem associated with defining forestry is the separation of the older profession of forestry, which was linked with survival, from technical forestry. Foresters, as technically-trained professionals, are responsible for guardianship. They are not trained either in managing the forests for human benefit or through human participation. Therefore, not only has people's participation in forestry suffered from lack of co-operation, but the concept of community participation is also lost in the wilderness of forestry professionals.

From the point of view of people, particularly the women who link forests with human survival, confusion has been created with the entry of an artificial form of forestry. Women are excluded from the forestry profession because of the assumption that guardianship is a masculine task and women are not capable of following the science of forestry. The art of forestry that decorates the landscape, primarily the isolated forest patches, homegardens and trees in the fields, is the forestry to which human life is attached. In this scenario, women and home maintenance is an integral part of forestry. The broad context of forestry includes a number of systems which consist of trees, either in the form of monoculture, or as part and parcel of the systems under concern. In Sri Lanka, this includes:

i   The state administered forestry sector,

ii   The traditional, informal sector of forestry.

Women's enrolment and contribution to these two broad categories vary according to the distinctive links they maintain.

The state administered forestry sector

The state forestry sector, administered by the Department of Forestry, is responsible for conservation of state forests, reafforestation, establishment of fuelwood plantations and community forestry programmes. Recently, it has extended its services to a number of Integrated Rural Development Programmes. Among these, the conservational task has been the main responsibility of the forestry department. This situation has displaced the traditional links that people had with the forests. Women's rights and access to forest conservational areas and plantations are not formally accepted as they are considered illegal creepers to the forests, while the male guardians are regarded as the foresters. The rigidity of the legislation, which never took into account sustainable uses of forests, has succeeded in eliminating women's links on a formal level. In reality, however, the law's weakness and professionals' incapacity to uproot women's links have meant that women continue to use forests in harmony with nature for subsistence.

One of the primary tasks of the forestry sector in Sri Lanka is reafforestation programmes. In these, no attempts are made to consult local communities regarding plantation programmes. The state makes decisions from the level of site and species-selection to harvesting. Community enrolment and inputs come in the form of hired labour and, in most cases, women are engaged as daily-paid workers in raising nurseries. State-sponsored forestry projects follow a top-down process, which filters from the central administration to regional divisions through a number of hierarchical levels. The entire plantation sector is under the authority of the Forestry Department and is meant to meet industrial timber requirements. The underlying concept is that forests are not meant for people, and programmes do not have to accommodate people's desires or entertain their participation. Thus, for instance, the species chosen for reafforestation mostly comprise Eucalyptus, Teak and Pinus. Forest fires on these plantations reflect the communities' feelings about this choice. According to people dwelling around the forests, Eucalyptus and Pinus provide homes for wild boar, which destroy crops and other vegetation.

The gathering of products, including deadwood, has been deemed a legal offence. The species chosen for these forest plantations do not yield the multiple products used by communities, so the only product to be gained from them is fuelwood. Thus, in times of acute scarcity, women creep into the plantations to collect deadwood from the forest floor. The forestry officials do not try to implement the law and prevent them because they have realised that women do not collect fuelwood destructively: they do not cut standing trees or saplings, but collect deadwood and branchwood (Wickramsinghe 1992e, 1993d and e). It is different with men, who tend to harvest wood illegally for sale. A cruel irony is evident in this situation: the people who are the best protectors of forests are now forced to exercise their roles on sufferance from those who are obviously still learning. Yet, still there are allegations that women are responsible for deforestation because of their fuelwood requirements. In fact, this segregation of forestry from people has become a threat to the forests since, practically, it has turned out to be difficult for the forestry officers to function without community co-operation.

Women in the community forestry programmes

A way out of the extreme segregation of the community from the forestry sector was sought through the strategy of community forestry. This strategy was adopted in 1982 to fulfil fuelwood needs through common sense and collective effort. Community forestry as a package was initially introduced in the Badulla District in 1982, and included programmes involving farmers' woodlots, community woodlots, demonstration woodlots and block fuelwood plantations. This package started with financial assistance from the Asian Development Bank (ADB). Since then, the programme has been extended to the districts of Nuwara-Eliya, Kandy, Matale and Batticoloa. Community enrolment has been initiated by distributing state lands, particularly the lands categorised as degraded and abandoned, on a 25-year lease agreement. Selected farmers were given about 0.2-0.4 hectares of land. The total area devoted to woodlots is about 632 hectares and about 1,861 families are enrolled.

A study conducted in Badulla district in 1991 (Wickramasinghe, 1991e), reveals that the farmer's woodlots programme has been attractive to families. As members of the family, women were expected to work in these woodlots, while 'farmers' were selected, accepting the 'male head' as the lease owner of the land. Out of 23 woodlots initiated in this process, 21 lots were listed with the names of these male heads and only 2 with names of women – both of whom happen to be widows. In this situation, women's ability to claim land rights is formally precluded. In delivering services and receiving training on woodlot management, the lease owners were privileged, so women were expected to depend on secondary information brought back by their husbands. Members of the 'village forest society' were dominated by males, bringing about a clear marginalisation of women. As a consequence, women's autonomy, particularly their self-determination and self-reliance in engaging in forestry activities for survival, has been violated. The officers of the state department not only demarcated the woodlots, but also assessed the community's needs. Instead of considering fuelwood as one of the by-products out of multiple outputs, the Eucalyptus introduced into these woodlots has been seen as a miracle 'fuelwood tree' recommended by the state.

In fact, what has happened is that community forestry has become a strategy for linking resource-poor families in adjacent villages with the forestry department in order to obtain labour. Over the last 10 years, these woodlots have fulfilled less than 10% of family fuelwood needs because, for women, standing trees were not the source of fuelwood and there has been no significant branchwood contribution. At present, these stands are stocks of straight timber that could probably be sold as round timber for transmission poles. In the absence of wood-based industries to consume the Eucalyptus wood, this will be the only way to sell timber. Women still have to walk long distances and collect dead roots of tea bushes for fuelwood.

The extent to which this approach integrated either the community or the women who bear the drudgery of collecting fuelwood is not clear. Similarly, the need for innovative community-centred activities and solutions to community-identified problems have remained at a theoretical level, and not permeated into practice. A remark of Hobley's (1987) is timely:

"As a notion, community forestry has not emerged from the village level, where the resource conflict occurs. It ignores the reality of this conflict and talks about forestry for local community development. A community of whom? A community of ruling class? A community of need?"

In fact, as Hobley concludes, the whole process ignores the issue of differential access to both natural and political resources. Numerous cases in Sri Lanka show that the whole programme suffers from the lack of community enrolment in the process of design, establishment and implementation. If the process was truly based on a community-oriented approach, women would have predominated in the process from the stage of design to the levels of implementation, utilization and evaluation.

None of these projects has taken into account the contributions that women could make to this sector. In most cases, 'women' are treated as a group of passive beneficiaries, who take the burden of finding the fuelwood for family survival. At the level of decision-making, this group of beneficiaries, comprising not only the poorest but also the well-to-do, have been subordinated to the men. As mentioned by women who attend community forestry meetings in Illuktenna in Badulla, they are the ones who possess minimum decision-making power. Theoretically, women are not alienated from the project. But their attendance is not meaningful because of male domination over decision-making and the inability of programmes to integrate the needs and problems of women themselves.

Often, women's participation has been seen as the engagement of their labour in raising nurseries for the Forestry Department, while the species identification was left to forestry officials. Women saw the undesirability of species like *Eucalyptus, Acacia mangiun, Pinus caribaea, Pinus patula, Gravillea robusta, Calliandra* and *Sesbasrica* as the main disadvantage. Whose trees are these? is a question unanswered by the officers. The women point out that, although these species are fast-growing and could be felled for fuelwood, their own needs are not restricted to fuelwood. They want species which produce multiple products, including fuelwood.

If community forestry projects are to succeed, they should begin in a community-based manner, beginning with the stage of project identification. For the success of the whole strategy, it is necessary to understand the community and how they practice forestry in their own contexts. The technological knowhow that foresters have could be integrated according to the wishes and desires of the community. Foresters could facilitate the process. Despite whatever is being said and done in adopting and elaborating the theory of community forestry projects, it remains a top-down process. The users' enrolment in conceptualisation, design and implementation of the programmes should not be subverted and dominated by the technology. Women could participate in designing the projects, using their experience, knowledge and needs to suit their own environment, if they are motivated and permitted to do so.

Women, social forestry and rural development projects

In about eight administrative districts, social forestry and agroforestry have been introduced as components of Integrated Rural Development Projects. In most districts, it is implemented in collaboration with the forestry department or in isolation. A large number of approaches have been adopted, primarily aimed at providing villagers with forest products and protecting the environment. This approach has three facets. First, there is the establishment of state authority over rights to the land and trees. At this stage, there is no intention of assuring community participation. The second facet is state-encouraged tree planting on lands leased to individuals or groups; the trees become the property of the planters. Village woodlots, farmers' woodlots and farmers' agroforestry plots fall into this category. The third facet of the programmes is private tree planting, where private landowners are encouraged to grow trees. This involves a number of approaches, including conservation farming with trees, homegarden programmes, minor export crops and interplanting of coconut.

In these programmes, women's concerns are still not accommodated during the implementation process. Subsistence needs are overlooked, yet women's labour is accepted by providing them with opportunities to earn a daily wage, equal to that of men, as nursery workers. But equality of wages does not mean that there is equitable distribution of the workload. Women are given labour-intensive tasks like raising and nurturing nurseries. No women have been recruited for supervisory work.

The main weakness of this range of elements is that none of them are formulated in a community-oriented manner. Neither the projects, nor women's participation in them, are determined according to women's felt needs or motivation. Rather than taking responsibility, they tend to work as requested by the officers. The programmes again become a state strategy for the recruitment of local labour without community consideration.

Among all these systems, the homegarden programme is the most popular with women, since it reflects their own choices and instincts. It is not surprising that the others fail to win their support within the short time-span expected by the officers. This would probably change if women were free to identify and design models rather than depend upon externally-designed.

Future programmes and the forestry sector

Increasing awareness of the consequences of deforestation resulted in the drawing up of the Forestry Master Plan of Sri Lanka, aimed at reforestation, conservation and protection. This plan was caused by the rapid loss of forest cover – from 2.9 million hectares to about 1.7 million hectares between 1956 and 1982, an annual reduction of about 42,000 hectares. Compared with the annual planting of about 7,600 hectares per year, these figures imply a widening gap between reforestation and deforestation.

The state sector cannot supply the 9.2 million tons of fuelwood and 10.9 million tons of industrial logs that are needed according to the estimates included in the plan. In examining current imbalances between consumption and sources of supply, it becomes clear that, in the absence of proper planning, the severity of the wood scarcity will be greater in future. At present, state efforts do not meet consumption needs, but there are still no plans to promote tree planting among farmers.

In the case of fuelwood, the non-forest wood biomass, which includes tea, rubber, coconut and homegarden products, is of great importance. This sector produces 50% of the industrial fuelwood and 80% of the domestic fuelwood used. In 1984, fuelwood comprised about 71% of the gross energy supply and 68% of it was consumed for household cooking. The pattern of household consumption varies among ecological regions, but it is the sole source of energy for 94-100% of households. The domestic sector alone consumes about 6 times the fuelwood required for the industrial sector of the country.

In the future, the competition from the industrial sector will increase. This is mainly because there is a strong tendency for the main sources of fuelwood to be absorbed into industry as raw material. For instance, if 49% of the wood energy supplied from rubberwood for the industrial sector is lost, the problems will become very severe. A situation similar to this has already occurred in Thailand. The increasing consumption of rubber-wood in industries as "white teak" will deprive the women of their supply. Industries are likely to purchase all the burnable sticks and wood, expecting women to depend on crop residues and the stuff which would not be purchased by the industrial sector.

The national plan does not address these problems. This omission seems to be due to limited concern with women's needs or household needs. If we are to depend on private sources as the main source of supply, then the forestry sector as a whole should be restructured to address local problems. This plan should have consulted women as contributors to the forestry sector. There is no viable alternative but to concentrate on the small-scale production sector and the sector of plantation agriculture, in which there is potential for tree integration.

Whatever the plans that the forestry sector have made in the past, none of them have fulfilled local expectations. In spite of millions of rupees from the Asian Development Bank (ADB), spent on community woodlots to meet rural fuelwood needs, no visible change has taken place in relieving the pressure on the rural poor. The fault lies not only on the wrong assessment of local community needs by the professionally-trained foresters, but on many other factors. One is the introduction of community forestry without baseline information. A further problem is the assumption that women will accept the selection of species and designs made by technical foresters because these will solve household energy problems.

Narrow perceptions prevent success in forestry programmes and masculine domination in forestry is a result of this narrow view (see Table 9). Women's participation has not gone beyond the grassroots level and the conflicting priorities of male decision-makers and women remain unresolved. To the women, farmers' fuelwood plots are timber stocks. Their expected outputs are not the outputs that women need. The 'science' of forestry differs from that of the culture and art of forestry continued by rural women.

Table 8 *Positions in the Forestry Department in Sri Lanka – 1990*

Job Title	Male	Female	Total
Conservator of Forest	1	–	1
Additional Conservator of Forest (1 - Administration) (1 - Operational)	2	–	2
Deputy Conservator of Forest	5	–	5
Divisional Assistants	5	–	5
Chief Research Officer	1	–	1
Research Officers	3	1	4
Divisional Forest Officers	14	1	15
Main Positions	26	2	28

Source Forestry Department (1990)

The discrepancies noted in Table 8 continue because the issues that cause the failure of forestry sector development are not properly examined. The priority research areas proposed by the Forestry Research Master Plan do not include community forestry and people's participation. Hence, in the second phase of the programme, the 'Participatory Forestry Project', the need for community research has been overlooked. Just as in the last century, research is still focused on genetic improvement, management of plantation forests, management of natural forests, conservation and protection of forests, non-wood forest produce and timber utilization. This focus shows an unwillingness to examine the constraints to the success of forestry, as well as the bias against focusing on conservational forestry. If forestry in Sri Lanka continues to be based on a masculine paradigm, the same mistakes of the past will only be repeated.

Women's Claim to the Role of Sustainable Development

Protecting the forest is termed guardianship. Men are regarded as the best guards for anything, hence women are excluded from the formal forestry sector. Yet, as we have seen, women have in fact been the most efficient users and managers of forest resources. Their activities are very economic in terms of time energy, and resources.

Figure 10 *Function of Forests in Survival Maintenance*

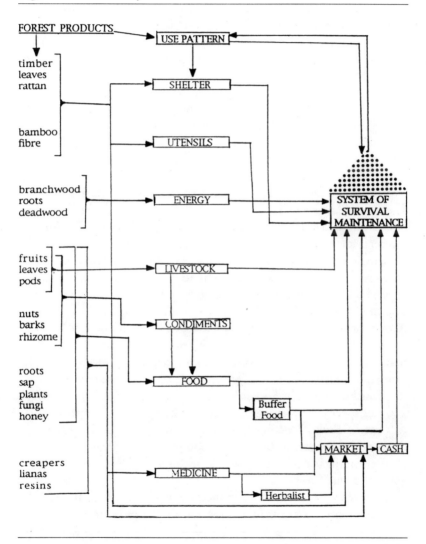

From the information in the above figure, it is evident that the wild food that women collect from the forests are also of medicinal value. Therefore, decisions regarding family diet are taken considering the nature of products that satisfy both food and medicinal needs.

Time allocation is very systematic. Gathering is done in a comprehensive way, to utilise one trip for many benefits. Wild edible leaves are collected at the same time as seeds, rhizomes, fungi, roots, resins, fruits and a few branches of dead wood. The habit of gathering varieties of products, irrespective of the quantity, increase their stocks at homes and assure availability throughout the year. Processing activities are lengthy – for instance, the extraction of oil out of kernels of Mee (*Madhuca longifolia*) and seeds of Wild-breadfruit (*Artocarpus nobilis*), so these do not begin until the fruit season is over, and the gathering, sun-drying and storing is complete. After seed-gathering is over, the extraction of oil from kernels begins, using labour-intensive indigenous technology.

Sufficient attention has not been paid to the non-timber uses of forests, as Clay (1988) has observed, partly because these products are often associated with traditional uses. The users of non-timber forest products are primarily women, because products of the forests gathered on the basis of household need. In Africa for instance, as much as 95% of traditional medicinal needs and herbal preparations depend on forest products (Eckong, 1979; Myers, 1982 and 1983). In assuring household food security, forest foods have played a vital role. According to Arnold *et al*, (1985); Campbell (1986); Gura (1986); Falconer (1990) and FAO (1989, 1990b) immense amounts of forest foods are gathered by traditional communities. Forest foods are primarily gathered for home consumption by the fringe dwellers of the Knuckles Range in Sri Lanka (Wickramasinghe, 1990b). Nearly 200 tree species are used by communities living on the periphery of the Adams' Peak Wilderness (Wickramasinghe, 1993i and j). Of all these, nearly 102 are medicinal and food products. Women are pioneers in the use of forest products inclusive of fruits, nuts, seeds and green wild leaves.

Because of their experience in the sustainable use of resources, women's control over the forests for non-timber products needs to be strengthened. It is not practically possible to isolate non-timber forest products from their users.

Women in farm and homegarden agroforestry

With the detachment of human survival from the forest, firmly-rooted forestry practices are still being used, since they provide subsistence and are environmentally sound. Homegardens reflect the desire for creating a liveable environment while linking the household with forestry in a realistic way. Trees maintained on the farmlands are scattered in their distribution and consist of naturally germinated ones as well as intentionally planted ones. These include jack (*Artocarpus heterophyllus*), coconut (*Cocos nucifera*), arecanut (*Areca catechu*), breadfruit (*Artocarpus altilis*), avocado pears (*Persea gratissima*), guava (*Psidium guajava*) and mango (*Mangifera indica*), all of which are purposely selected, as well as the ones which have entered naturally. Women determine the location of these trees. Newly promoted food-producing species, particularly the early fruitening varieties, are planted in the homegardens, while a few others are maintained in the farmlands which, meant for crop monoculture, are primarily controlled by males. Often, women select the best varieties among the ones available to propagate in homegardens.

Women tend trees which produce multiple products. The ideal combination is determined by the multiplicity of the products. Food, fuelwood, fodder and organic residues for mulching are the products they are concerned with, and so the species which are best deemed either satisfy all these needs or a number of them. If one ranks species accordingly, it becomes clear that jack (*Artocarpus heterophyllus*) and coconut (*Cocos nucifera*) are preferred, and even the fruit-producing ones like mango (*Mangifera indica*), avocado (*Persea gratissima*) etc. are listed below them. Any production system based on multiple tree species enables women to prepare a well balanced diet. Fruits like jack, or breadfruit (*Artocarpus altilis*) are often cooked as a main dish or as a substitute for rice during shortages. During the fruit season, women store the excess for the off-season by means of processing and preserving. In this way they are able to ward off the risks of food shortage. They protect *Madhuca longifolia* which grown along the boundaries of the paddy tracts for the purpose of collecting kernels to extract oil. This oil is often a substitute for coconut oil.

Women's tasks in homegardens, which are the most widely-seen tree-based system, are not restricted to enriching the composition of homegardens, but also include the transfer of biomass. The organic manure gathered in the homegarden is accumulated in pits for decomposition and mixed with animal dung to enrich the soil. Through this process, women physically integrate forestry, farming and animal husbandry for the sustenance of the homegardens and subsistence crop production.

The failure to appreciate and understand the tasks performed by women and their traditional tree-use practices has resulted in the failure of state-intervened tree farming. The priority given to timber production for the industrial sector, which is not women's primary concern, has left indigenous forestry skills untapped. Similarly, in agriculture, by giving priority to food-grains, the role of trees in solving food problems has been overlooked (Wickramasinghe, 1990e). Lack of incentives to promote agro-based industries which offer a range of potentials has caused farmers to remove trees from their farmyards. As is emphasised by Stephard (1985), inability to understand the extent to which women's activities relate to resources and their management is one of the persistent failures in development. Women's links to resources need to be recognised beyond their engagement as users. They determine the subsistence economy interwoven with forestry and subsistence remains their priority in crop production, forestry and land use.

With regard to fuelwood, the neglect of multiple uses of trees through isolating energy from other uses, offers us very limited hopes. Similarly, concentration on seasonal crops which are grown in monoculture has led to a greater number of adverse effects because it increases the dependency of the rural poor on the market. If the poor have to depend more and more on the market, then they have to be sure of a reliable and regular income to increase their purchasing capacity. A deepening poverty situation, which is linked with deteriorating household food security, has partly contributed to the retrieval of trees. But tree products that offer food substitutes for the poor are becoming scarce. In this context, the promotion of trees in the farmland should be treated as a way of stabilising the environment while meeting survival needs. Development strategies that exclude women's needs, knowledge and problems reflect the unequal distribution of resources, the imbalance between subsistence and commercial products and a gender bias in delivering assets and stimulations.

The outcome of this situation is that women's contributions to forestry are concealed behind their domestic tasks because their forestry-related activities are also related to home-maintenance. Although the homegarden lands are not legally owned by women, their control there is greater than in any other production system. Their success in this area can be measured by their ability to maintain the system without external support. Sustainability of forestry can learn a great deal from homegardens. In the formal forestry sector, it is believed that continuity of forestry depends on investment in cash rather than on people's commitment. Yet, in many rural areas, women's commitment to having a functioning home is being extended to homegardens, as well as to forestry.

If forestry is regarded as a means of uplifting the standard of living of people, resources should be allocated for this purpose. But the sectors which have control of the funds are not interested in this. In promoting forestry for local communities, we expect more from people: their time, skills, knowledge and experience. Instead of accepting their labour on the grounds that they are the beneficiaries, women need to be recognised as contributors who build up forestry in a sustainable manner, linking survival needs with expected outputs.

Although foresters' attention is mainly focused on scientific research towards the mass production of timber, it is known that the generations-old experience of local people is ignored. This may be due to the difficulties of incorporating practices which are bound up in a complex web of human culture. However, their ways of collecting germ- plasm, sharing of knowledge, information and resources through social contacts, and provision of products from the neighbourhood are of primary importance.

Conclusions

There is a strong segregation between the state forestry sector and the informal sector which includes all the agroforestry systems managed by people. State control over the formal sector and community control over informal forestry is another form of this segregation. Women, as an equal half of humanity, are heavily engaged in forestry and especially in the sustainable use of forests.

The current situation shows that even if the forestry sector remains focused on timber supply, the widening gap between timber needs and sources of supply literally cannot be narrowed down through the state sector. The scarcity of land for establishing tracts of tree monoculture on the one hand, and the control of other sectors over the rest of production systems on the other, are obstacles to the expansion of the state sector. It is therefore necessary to shift from tree monoculture to multiple tree production.

At the same time, seasonal crop monocultures should accommodate trees. In terms of demand, the most emphasised aspect is timber requirements. This is, in fact, not met by forest plantations alone. Nearly one-half of timber needs are met by rural people, among whom women play a prominent role. It is necessary to address industrial wood requirements as well as the need for other forest products, from the level of household to the national level. At present, household survival needs and food security are concerns outside the sphere of interest of the forestry profession. Similarly the need for integrated land-use are also beyond their interests. If forestry is organised to meet multiple needs and formulated in an integrated manner, community enrolment will become strong. With population pressure increasing and resources dwindling, such a policy change would present one way out of the country's current impasse.

Another question that needs to be raised here is whether it is possible to fit forest and farm trees into a commercial economy. It should be noted that forests and trees contribute many products which could be consumed, sold directly or preserved and then sold. For instance, the increasing demand for fuelwood in the urban market and for rural industries offers a market for rural dwellers, if they could produce fuelwood in excess. As the FAO (1989a) argues, tree products could supplement farm production, filling in seasonal shortfalls in food and income and providing a buffer during periods of hardship. Income from the products could easily supplement the household budget and provide multiple products for the household, including fuel and food. Rural women consider trees and forests as buffer food sources and security against hardships and drought.

Genuine forestry in Sri Lanka is in fact in the hands of small-scale farmer families, in which women are the most capable of contributing to the forestry sector as a whole. This is due to their primary concern over trees and the traditional tree-use practices maintained by them, and also due to their direct involvement in household farm operations.

The major weakness of the formal forestry sector lies in its inability to grasp community needs, tree-use practices and potential contributions. Attempts should be made to reduce these gaps through a community-oriented approach to forestry. Although community-oriented forestry has been accepted, the formal forestry sector, which controls resources, is not suitably organised to accommodate it. It is not realistic to expect local women to provide their labour for the sake of development, when they have neither control over resources nor any place in the process of making forestry decisions.

CHAPTER 6

Priorities and policy issues

From the information gathered for this study, it is clear that forestry should become a strategy to establish a sustainable life-support system linked with agricultural and energy policies. The future of forestry is in the hands of small-scale farm families who traditionally managed the land in an integrated manner. The penetration of crop monoculture has distorted the economic and environmental role played by farm families as well as by trees. In redressing the problems created, forestry should be regarded as development strategy. Here, foresters will have to mobilise local resources and work in collaboration with the community. Women should be regarded as the focal point. Multiplicity of forest products and renewability of resources are the key considerations in this. In reformulating policies towards meeting survival needs, the following points deserve attention.

i In the process of forestry development, women's roles are essential. As the greatest non-timber users of forests, decision- making should be in their hands. The time they spend gathering forest products should be regarded as a contribution to the sector. The gender-disparity in attitudes towards trees i.e. women's lasting interest in tree products and men's interest in timber obtained through destructive harvesting, can be reconciled through the promotion of tree planting. Male collaboration in making women free to accept the challenge is as important as that of forestry desires to tap women's engagement. Thus, the tasks done by women should not be taken as biological commitments, nor should women's involvement be justified on eco-feminist theories that too easily link women with nature. Attention should be drawn to tree-based cultures nurtured by women.

ii Trees must be integrated with other land-use practices, avoiding possible conflicts with the planting of crops. This integration should begin at the level of state administration, where all the sectors like forestry, agriculture, animal husbandry and export agricultural crops should be combined. This restructuring should be founded on traditional multiple land-use policy. Although all these measures could be accommodated in one basket, the choice of selection must be left to the households, in which women would come up the with best choices because of their experience in multiple land use and integrated farming.

iii Two other conditions are required for the efficiency of this type of multiple and integrated approach. First, the prevention of deviation of the sectors through the introduction of an integrated theme in delivering services. The second is the reduction of male domination in the professions of forestry, agriculture and animal husbandry, which has remained an obstacle to understanding gender issues in farm forestry, home gardening and farming systems. Male-dominated services have a negative impact on women's participation. This may be due either to the difficulty of attracting women who are busy with a number of tasks, or to the women's reluctance to voice their needs and problems and discuss them freely with male officers. Whatever the case, the promotion of women as professionals in these sectors would help to address the problems in an efficient manner. Although, women are not technologically-trained their traditional technologies are much more appropriate to their specific environment.

iv At the same time, the state sector as a whole should appreciate women's contribution. So far, women's participation is expected as beneficiaries. Forestry has been planned by the foresters on the assumption that their decisions are appropriate to women. Yet, it is women who have experience of the most desirable measures inherited from their ancestors. Participatory research is necessary to realise and recognize women's contribution from the lowest level of the social strata to the level of policy planning. Unless research can document women's roles and their experience, technology will exclude the knowledge of traditional forestry. Women often posses a unique indigenous knowledge about tree species, locations for germ-plasm, ecological niches in which each species survives best, low-cost conservational technologies, multiple uses of tree products, technologies of preparing, processing and preserving products and technologies in multiple land-use. These aspects are seldom taken into consideration in development planning. By sharing them, policy planners would be able to formulate more appropriate, desirable and adoptable technologies. It is important to realise that by tapping women's knowledge alone, it would be possible to eliminate several years of experiments and trials.

v Hence, a participatory process is desirable. Whose participation is needed, how and what is to be done are the important issues. People's participation has been discussed in detail by researchers, field extension workers, NGOS and others. According to Chambers *et al*, (1983) and Chambers and Ghildyal (1985), it is the first step in agricultural development. As has been pointed out by Hiemstra *et al* (1992), farmers are the ones who should judge what is needed. As the main carriers of water for domestic needs and the principal moulders of families' hygienic habits, women's involvement in decision-making is of critical importance (Srinivason, 1990).

Over the years, many efforts have been made to ensure people's participation in projects and programmes with limited success. Success depends largely on which members of the community are being approached. In most cases, women are regarded primarily as the beneficiaries and not approached with the interest of mobilising them. What is more serious is that they are merely regarded as cheap labour. Participation in forestry encompasses a lengthy process, starting with women's selection of species and germ-plasm, to the stage of managing, harvesting, processing and marketing. It is not enough simply to enable women to take part, but it is also necessary to initiate their involvement from the beginning, based on their knowledge and needs. Foresters should play the role of facilitators. They should initiate a process of forestry development that is suited to local habitats and acceptable to local inhabitants.

In most cases, when extension services go to the community, women are ignored. These services often fail to convince them of the desirability of the strategies being promoted, which often conflict with their own experience. To women, many of the modern designs are artificial. Sloping Agriculture Land Technology (SALT) is a clear example. For local women, it has violated the contour-farming that they have practised in hilly areas. By establishing Gliricidia on lines along narrow strips of land in SALT systems, the multiplicity of contour-farming has been ignored. In many situations, the most desirable strategies are the ones practised by women, although such talents do not surface as they have no control over the flow of external interventions, which comes with a package of subsidies, services and inputs.

vi Women must be specially targeted from the stage of project formulation. It is necessary to analyse their needs and problems on the one hand and their knowledge, technologies and experience on the other. At the same time, it is important to identify their wishes in tree farming. A whole series of forestry activities should be oriented to them. They need to be regarded as key actors in the selection of species, planning of designs and control of the systems that deliver resources. We must allow women to decide what they want and how to fulfil their survival needs, minimising the use of time and energy. This is of vital importance to rural women, mainly because there are no alternatives for tree biomass.

vii The participatory approach in the forestry sector should be applied to identification of species, site-specific ideotypes and designs and ecological niches for the growing of specific species (Wickramasinghe, 1992e). Women's designs are biologically diverse because neither trees of same ideotypes nor the same tree is grown everywhere. The fundamental issue of their future prospects could be combined with these, and women could be encouraged to forward their own views. This theme should be adopted in the best manner to encourage women to take part in making decisions on planning multiple land-use systems and to integrate their contributions in a practical way to achieve the sustainability of development.

viii The status of rural women is not equal to that of their male partners in terms of literacy, education, freedom of mobility and positions in community organizations. This status should not be attributed to the innate passivity of women. Women have not been given incentives to express their hidden abilities, skills and experience.

It is important to draw on their survival strategies that combine needs with available resources. Their priority for household well-being and domesticity is often taken as an excuse to marginalize women in the process of development. If development is meant to promote a liveable environment and improve well-being, the women who have prioritised survival needs and well-being must be targeted first. Continuity of the systems largely relies on the women of the present generation, who will transmit their knowledge to the next.

ix In forestry, the failure to understand women's tree-use practices and management strategies has allowed a hierarchy of male domination. Women should be accommodated in each strata of this sector, from the level of policy planning to the level of implementation. Unless this representation is encouraged, accepted and supported, significant changes can hardly be expected.

x Many of women's forestry tasks are based on tradition. With the changes in the economy, where men are absorbed into the commercialised sectors, women must also be given opportunities to make similar adjustments. We cannot claim women's participation without soliciting men's will. This is also to be initiated at the level of their survival unit, which is the family.

xi These changes alone will not be adequate to promote women in forestry. It is necessary to reorganise the arrangements and remove obstacles to their active participation. In the absence of women's organizations, it is difficult to contact women at the lowest strata of society in an official or professional manner. This is important in drawing the attention of extension services and making available credit and training. Common or group requests are more sound than individual voices. Women's organisations should be strengthened. The same organisational structure could be used in delivering extension information to women. In their own community settings, women are the agents of information-sharing. Those who have no access to information cannot be expected to share the benefits of services with those who have it. This also applies to training. In reality, the benefits of training go to the category who receives information. This category often consists of the better-off and males.

If 50% of humanity is not incorporated, and trust is placed in the same 50% to take part in operationalization, the consequent imbalance in development thinking will become a constraint to the success of development efforts;

xii Tenurial limitations should be lessened in favour of women. The women who contribute to family food production should be equally eligible for receipt of training, inputs, subsidies and other facilities. If bonds of family form the basis of survival, women should not be prevented from receiving the benefits of services delivered on family property. The other situation, in which women are not engaged to plant trees on common land, should not be interpreted to mean that women have no interest in forestry. They abstain from planting since they will have no rights to claim the outputs. If the communal areas are to be reafforested, then the same labour must be able to claim the products. This type of policy transition is equally important in motivating women to plant trees. However, as discussed by Fortman (1985) the tree tenure factor needs much attention in promoting agroforestry.

xiii State policies should be reformulated to prioritise tree-farming on private lands. A scheme of inputs must be introduced for this purpose, as for many other agricultural crops. At the same time, bio-technologies should be used to promote trees which suit locations and products. Processing and preserving technologies are also to be combined. Tree-farming must be promoted as an avenue of survival. The aim should be diversification of the sources of income and maximisation of the production levels of the small-scale farm operator, whose land and capital is limited. Tree crops and livestock should be combined. But predetermined models, without the participation of the people concerned, will not be successful. Control of environmental degradation and promotion of the resource base are to be appreciated as benefits brought by such systems.

xiv Women do not find tree-growing a new practice. Trees were part and parcel of their survival. Direct benefits, i.e. tree products, and indirect ones such as a healthy environment, are known to women. We talk about women's forest management practices within a theme of non- timber forest products. The loss of such systems are considered by them as the loss of their means of sustenance. In order to restore the sustainability of survival systems, efficient policy measures are needed to revive such strategies. The state must be prepared to subsidise the cost of women's time and energy, but seedlings should not be chosen for specific locations based on what is available to the forester. These should be decided by the people concerned. Agricultural and forestry policies should not be in conflict. These policies must be combined in a practical manner and incentives must be given in favour of the land users.

xv Under the circumstances prevalent in Sri Lanka, the necessity for stimulating community forestry – not in mass systems or in the form of tracts of trees, but within a theme of multiple and integrated land-use – is well understood. By focusing attention on women, the attempt is not to add another responsibility to women who are already struggling to satisfy survival needs. In the first place, under the commercialised economy which is penetrating into the rural environment, men often migrate or find employment or casual paid work. In such situations, women are the ones who undertake farming. In addition, with the consequent reduction in soil productivity, men of such farm families hire their labour for cash returns, which is equally important for meeting family needs. Therefore, women remain the best target group to be enrolled in forestry.

Trees should be treated as a crop. During the season of cultivation which follows the occurrence of rainfall, selected tree seedlings can be raised by women and different species can be distributed among them. This should not be done as a tree-planting campaign within a specific day or a year. A number of seedlings of the species desired by women could be given initially with state assistance and continued over a number of years. By doing so, it would be possible to gradually invest in forestry the time which is often spent in collecting fuelwood from long distances in the forest. This absorption into the forestry sector of time and energy spent on gathering tree products should expand over a period of time.

xvi   Some adjustments should be made within the sphere of the household in order to reduce the heavy workloads of women. Men, as partners in household survival systems, could, of course, share many of the tasks done by women. Then the latter would be relieved, at least for one hour or so, from other tasks, and such time would be profitably utilized in farm forestry. Women's efficiency in doing multiple tasks and nurturing trees would enable them to combine reforestation and environmental rehabilitation, if other requisites are provided. In any case, these are part and parcel of farming practices.

xvii   Quite simply, trees remain central in providing multiple needs and women play the key role in supplying survival needs and utilising the products. Whatever the concepts we may adopt in forestry development, its ability to meet future generations' needs depends on the acceptability and adoptability of strategies to local conditions. Local conditions include both the physical and human resource-base. Among local people, women are the leaders in utilising forest products efficiently, hence forestry, as a sustainable development measure, must be linked with current survival as well as efforts to improve the future.

In this context, the role of women is much broader than that of being gatherers and nurturers. They could become the agents of forestry extension, because they often share knowledge, materials and information through their social contacts. Unless we mobilise such resources, forestry extension will be limited to delivering external strategies to the top strata of local people. In a broad sense 'forestry is about people', and this includes all the production activities, their behaviour and consumption. Technical aspects related to modern forestry sectors could bring together people through state intervention with the intention of improving their own lives. In fact, by supporting and strengthening women's roles in the forestry sector and in society, it is possible to contribute to the effectiveness of development intervention, while improving women's status, the well-being of their families and the environment.

References

Abramovitz, J.N. (1991). *Investing in Biological Diversity Countries.* U.S. Research and Conservation efforts in developing countries. World Resource Institute.

Adams, W.M. (1990). Greening of develop trend. *Green Development. Environment and Sustainability in the third world.* Routledge – London and New York.

Agarwal, B. (1986). *Cold Hearths and barren slopes. The woodfuel crisis in the third world.* Zed Books, London.

Arnold, T.H. *et al* (1985). Khosian food plants: Texa with potential for future economic exploitation. G.E. Wickens *et al* (eds.), *Plants for arid lands.* Allen & Unwin, London.

Berry, W. (1977). *The unsettling America – culture and agriculture.* Sierra Club Books. 228 pp.

Bhasin, K. *et al* (1991). *Our indivisible Environment.* A report of the FAO-FFHC (FAO Freedom From Hunger Campaign and Action for Development). Workshop in South Asian Environmental Perspective. Bangalore, October 1-7, 1990, New Delhi; design and print.

Blaikie, P. and Brookfield, H. (1987). Defining and debating the problem. In *Land degradation and Society.* Blaikie, P. and Brookfield, H. (eds.) pp. 1-27. Methuen, London and New York.

Bruntland, H. (1987). *Our Common Future.* Oxford University Press, Oxford. For the World Commission on Environment and Development.

Budd, W.W., Duehhart, I., Hardesty, L.H. and Steiner, F. (eds), (1990). *Planning Agroforestry.* Elsevier, Amsterdam, Oxford, New York.

Campbell, A. (1986). The use of wild food plants in drought in Botswana. *Journal of Arid Environment,* 11(1): pp. 81-91.

Cecelski, E. (1985a). *The Rural Energy Crisis, Basic Needs and Women's Work: What can Donors Do?* (Paper for women in Development Expert Group Meeting DAC/OECD); Paris, 29-31 Jan.

Cecelski, E. (1985b). *The Rural Energy Crisis, Women's Work and Basic Needs: Perspectives and Approaches to Action.* Geneva. International Labour office.

Cecelski, E. (1987). *Energy and Rural Women's work: Crisis, Response and Policy alternatives.* International labour review.

Cecelski, E. (1992). *Women, energy and environment: New directions for policy research,* International Federation of Institutes for Advanced Study (IFIAS), Toronto, Canada.

Chambers, R, Saxena, N.C. and Shah, T. (1989). *To the Hands of the Poor. Water and Trees.* Oxford and IBH Publishing Co. Pvt. Ltd. New Delhi. Bombay. Calcuta.

Chambers, R. (1983). *Rural development: putting the last first,* Longman, Harlow.

Chambers, R. and Ghildyal, B.B. (1985). *Agricultural research for resource for farmers: the farmer-first-and last model.* Agricultural Administration and extension. 20, pp. 1-30.

Chavangi, N.A. Engelhard, R.J. and Jones, V. (1988). *Culture as the basis for implementation of self-sustaining Woodfuel Development Programmes.* In: *Whose Trees? Proprietary Directions of Forestry.* (ed). Fortman, L. and Bruce, J.W. pp. 253-254. Westview Press. Colorado.

Chinwuba, Obi, S.N. (1988). Women's Rights and interests in Trees (Nigeria). 240-243. In *Whose trees? Proprietory dimensions of Forestry.* Fortman, L and Bruce, J.W. (ed).

Clay, J.W. (1988). Indigenous peoples and tropical forest: Models of land use and management from Latin America, *Cultural Survival,* Report No. 27.

Conroy, C. and Litvinoff, M. (eds.) (1988). Greening of development. *The greening of Aid: Sustainable livelihoods in practice.* Earthscan, London.

Corea, G. (1975). *The Instability of Export Economy.* Marga Institute – Colombo.

Dankelman, I. and J. Davidson. (1988). *Women and environment in the third world: Alliance fo rthe future,* Earthscan Publications Ltd., London in association with IUCN.

De Beer, J.H. and Mcdermott, M.J. (1989). *The economic value of non-timber forest products in Southeast Asia.* Netherlands committee for IUCN. Amsterdam, The Netherlands.

Digerness, Turi Hammer, (1977). *Wood for fuel. The energy situation in Bara, the Sudan.* Mimeo, Department of Geography, University of Berger, Norway.

Directorate General for International Cooperation (DGIC). (1990). *Women, Energy, Forestry and Environment.* Ministry of Foreign Affairs, The Netherlands.

Draper, P. (1975). 'King Women: Contrasts and Sedentany contexts', In: Rayna R. Reifer (ed.), Toward and Anthropology of Women, New York and London: Monthly Review Press.

Eckholm, E.P. (1975). The other energy crisis. *World Watch Paper No. 1,* World Watch Institute, U.S.A.

Eckholm, E., Gerald Foley, Geoffrey Barnard and Lloyd Timberlake (1984). *Fuelwood, the energy crisis that won't go away.* An Earthscan paperback, London and Washington D.C.

Ekong, D.E.V. (1979). African Medicinal Plants under the Microscope, *Courier* (UNESCO Monthly), May 1979: pp. 17-18

Falconer, J. (1987). *Forestry and diets,* FAO, Rome.

Falconer, J. (1990). Hungry season food from the forest. *UNASYLVA,* Vol. 41, No. 160, pp. 14 19.

FAO. (1978). *Forestry for local community development.* FAO-Forestry Paper no 7. Rome. Food and agriculture organization of the United Nations.

FAO. (1981). *Map of the fuelwood situation in the developing countries.* FAO, Rome.

FAO. (1983). *Rural women, forest outputs and forestry Projects.* FAO, Rome.

FAO. (1984). *Yearbook on Forestry Products* (1982). FAO-Forestry Statistics Series, Rome.

FAO. (1985). *Tropical Forestry Action Plan.* Committee on Forest Development in the tropics. FAO, Rome.

FAO. (1987). *Restoring the Balance: Women and Forest Resources.* FAO, Rome.

FAO. (1989a). *Household ford security and forestry: an analysis of socioeconomic issues.* FAO, Rome.

FAO. (1989b). *Forestry and food security,* FAO, Rome.

FAO. (1990a). *Women and Forestry.* Tenth Session, 24-28 September 1990. FAO, Rome, Italy.

FAO. (1990b). *The major significance of minor forest products,* FAO, Rome.

FAO. (1991). *Women's role in dynamic forest-based small scale enterprises,* FAO, Rome.

FAO. (1993). *Women and forestry, Gender issues in the Asia Pacific Region,* Secretariat note. Presented at the 15th Asia- Pacific Forestry Commission, Colombo, Sri Lanka.

Fisher, R.J. and Gilmour, D.A. (1990). Putting the community at the centre of communnity forestry research. In: M.E. Stevens, S. Bhumibhamon and H. Wood (eds.), *Research Policy for Community Forestry, Asia-Pacific Region,* RECOFTC, Bangkok, Thailand, 5: 73-80.

Floor, W.M. (1977). *The energy sector of the Sahelian Countries,* Mimeo, Policy Planning Section, Ministry of Foreign Affairs. The Netherlands. April.

Ford-Robertson and R.K. Winters (1983). Terminology of forest Science, Technology, Practice and Products, *Society of American Foresters,* Washington, D.C., U.S.A.

Fortman, Louise (1986). Women's Role in Subsistence Forestry. *Journal of Forestry.* 84 (7): 39-42.

Fortman, Louise and Bruce, John, W. (1988). *Whose trees? Proprietary dimensions of forestry.* (Rural Studies sereis). Westview Press, Colorado.

Fortmann, Louise (1985). The tree tenure factor in agroforestry with particular references to Africa. *Agroforestry Stem 2;* pp. 229-251.

Foster, T. (1986). A common Future for women and men (and all living creatures). A submission to the *World Commision on Environment and Development,* EDPRA. Consulting Inc., Ottawa, Canada.

Gura, S. (1986). A note on traditional food plants in east Africa: Their value for nutrition and agriculture. *Food and nutrition,* 12(1): pp. 18-26.

Harrison, P. (1987). Greening of Development, *The greening of Africa: breaking through in the battle for land and Food.* Paladin, London.

Hiemstra, W.C. Reijntjes and Erik van der Werf. (1992) *Let farmers judge: Experiences in assessing the sustainability of agriculture.* Intermediate Technology Publication, ILEIA.

Hombergh, H. v.d. (1993). *Gender, Environment and Development; A guide to the literature.* Institute for Development Research. Amsterdam (InDRA) Amsterdam, The Netherlands.

Hobley, M. (1987). *Involving the poor in forest management. Can it be done?* The Nepal Australia Project experience. p. 6. Overseas Development Institute. Network paper 5C, October 1987, Nepal.

Hoeksema, J. (1989). *Woman and Social Forestry.* Bos Document 10. Bosbouw Ontwikkelings Samenwerking Amsterdam/Khartoum.

Hoskins, M. W. (1979). *Women in forestry for local community development – a programming guide.* The office of women in Development, USAID, Washington, D.C.

Hoskins, M. (1979). *Women in community forestry. A field guide for project design and implementation.* p. 45. FAO, Rome.

Hoskins, M. (1980). *Community Forestry Depends on Women.* Vnasylva, 32 (130): 27-32.

Hoskins, M. (1982). Community Management of Natural Resources. In: *Women in Natural sources: an international Perspective.* ed. by Molly Stock et al, Forestry wild life and Range Sciences, U. of Idaho, Moscow, U.S.A. July.

Hoskins, M. (1983). *Rural women, Forest Outputs and Forestry Projects.* Discussion Paper No. MISC/83/3, FAO, Rome.

Hughart, D. (1979). *Prospects for traditional and non-conventional energy sources in Developing Countries.* World Bank Staff Working Paper No. 346. World Bank, Washington, D.C. July.

Huxley, P.A., D.E. Rocheleau and P.J. Wood (1985). *Farming Systems and agroforestry Research in Northern Zambia. Phase 1 Report: Diagnosis of kind use problem and Research indications.* Nirobi: ICRAF.

ILO. (1966). Maximum permissible weight to be carried by one worker. *Report to the International Labour Conference.* Fifty first Session. Geneva, ILO.

ILO. (1985). *Energy and Rural Women's work. Vol 1:* Proceedings of a preparetory meeting on energy and Rural women's work. 21-25th October 1985. ILO, Geneva.

ILO. (1987). *Linking Energy with Survival: A guide to energy, Environment and Rural women's work.* ILO Geneva.

IUCN. (1980). *The World Conservation Strategy. International Union for Conservation of Nature and Natural Resource.* United Nations En Programme, World Wildlife Fund, Geneva.

Kumar, S.K. and Hotchkiss, D. (1988). *Consequences of Deforestation for women's time allocation, agricultural production, and Nutrition in Hill areas in Nepal.* International Food Policy Research Institute.

Leslie, A.J. (1987). Foreword. In: Westoby, J. (1987). *The purpose of forest – follies of Develeopment.* Basil Blackwell Ltd., Oxford.

Lundgren, B. (1987). Introduction. *Agroforestry systems.* 1: 3-6. Ministry of Lands and Land Development (1986). *Forestry Master Plan for Sri Lanka.* Ministry of Lands and Land Development.

Mok, S.T. (1990). Predicted future directions in the management of forest. In: Research policy for community forestry Asia-Pacific Region, M.E. Stevens, S. Bhumibhamon and H. Wood (eds.), pp. 16-24, RECOFTC, Bangkok, Thailand.

Myers, N. (1982). *The use of wild plants for traditional medicine in Africa.* Report to the World Health Organization (WHO), Geneva, Switzerland.

Myers, N. (1983). *A wealth of wild species: storehouse for human welfare.* Westview Press, Boulder, Colorado, U.S.A.

Myers, N. (1986). Tropical deforestation and a Megaextinction spasm. In: Soule, M (ed): Conservation biology – The science of scarcity and diversity, Sinaker Associates, Massachusetts.

Nagbrahman, C. and Shreekant, S. (1983). Women's Drudgery in firewood collection. *Economic and Political weekly.* January.

Panayotou, T. and P.S. Ashton. (1992). *Not by timber alone: Economics and ecology for sustaining tropical forests.* Island Press, Washington D.C.

Perera, N.P. (1984). Natural Resources Settlementa and land Use. In: *Ecology and Biogeography in Sri Lanka,* edited by C.H. Fernand. Dr. W. Junk Publishers. The Hague.

Pieris, R. (1956). *Title to land in Kandyan Law.* Sir Paul Peiris Felicitation Volume: 92-113.

Postel, S. and Heise, L. (1988). World Watch Paper. 83, *Reforesting for earth,* World Watch Paper, 83: World Watch Institute, Washington D.C.

Poulsen, G. (1978). *Man and trees in tropical Africa.* Publication no 101C. International Development Research Centre. Ottawa, Canada.

Raintree, J.B. (1987). The State of Art of agroforestry diagnosis and design. *Agroforestry Systems,* 9: 219-251.

Raintree, J.B. (1990). Agroforestry Diagnosis and Design: overview and Upgrade. pp. 33-58. In: *Planning Agroforestry.* W.W. Budd, I. Duchhart, L.H. Hardesty and F. Steiner (editors), Elsevier Science Publishing Compory Inc. New York.

Raintree, J.B. (1991). *Socio-economic attributes of tree and tree planting practices.* FAO, Rome.

Raintree, J.B. (1992). Community-based tree improvement: a new series of activities, beginning with the Artocarpus network. In: *Research on farmer's objectives, for tree breeding* (eds. J.B. Raintree and D.A.Taylor. pp 70-74 F/FRED Bangkok).

Rao, Y.S. (1990). Community Forestry Research: An Asia-Pacific Overview. In: M.E. Stevens, S. Bhumibhamon and H. Wood (eds.) *Research Policy for Community Forestry,* Asia-Pacific Region, RECOFTC, Bangkok, Thailand, 5: 10-15.

Rao, Y.S. (1992). *Tropical forestry:* an Asia-Pacific Perspective. In: *MPTS in Sri Lanka, Research and Development,* H.P.M. Gunasena (ed.), pp. 1-17.

Reven, P.H. (1988). *Our diminishing tropical forests, in Biodiversity* (ed. E.O. Wilson and F.M. Peter), National Academy Press, Washington D.C., pp. 119-122.

Risseeuw, C. (1988). *The fish don't talk about the water: gender transformation, power and resistence among women in Sri Lanka.* Brill, Leiden, The Netherlands.

Rocheleau, D.E. (1985). *Land-use Planning with rural farm households and communities.* (Participatory Agroforestry Research). Agroforestry. International council for Research in Agroforestry. Nairobi. 76 pp.

Rocheleau, D. (1986). Criteria for re-appraisal and re-design: Intra-household and between household aspects of FSR/E in three Kenyan Agroforestry Projects. In: *Selected Proceedings of the annual Symposium on Farming Systems Research and Extension.* Ect. 7-14, 1984 – C.B. Flora and H. Tommacet, eds., pp. 456-502.

Rocheleau, D. (1987). Women, trees and tenure: Implications for agroforesty Research and Development. In: J.B. Raintree (ed.). *'Land, trees and Tenure':* Proceedings of an International Workshop on Tenure Issues in Agroforestry, ICRAF, Nairobi.

Rodda, A. (1991). *Women and the Environment,* Zed Books Ltd, London and New Jersey.

Saka, A.R., Bundenson, W.T., Mbekeani, Y. and Itimu, O.A. (1990). Planning and Implementing agroforestry for small-holder farmers in Malawi. pp. 247-267. In: *Planning Agroforestry,* W.W. Budd, I. Duchhart, L.H. Hardesty and F. Steiner (editors), Elsevier science publishing company Inc. New York.

Saxena, N.C. (1987). Women in forestry. *Social Action. Vol. 37*, April-June 1987. pp. 150-162.

Scherr, S.D. (1990). The Diagnosis and Design Approach to Agroforestry Project Planning and Implementation: Examples from Western Kenya. pp. 132-161. *Planning Agroforestry*. W.W. Budd, I. Duchhart, L.H. Hardesty and F. Steiner (editors), Elsevier Science Publishing Company Inc. New York.

Schrijvers, J. (1985). *Mothers for Life*. Motherhood and marginalization in the North Central Province of Sri Lanka. Delft: Eburon.

Seneviratne, E.W. (1982). National Forestry Extension Programme. *The Sri Lanka Forester*, vol. 15 no. 3 and 4. 113-117.

Shiva, V., J. Bahdyopadhyay and N.D. Jayal. (1985). Afforestation in India: Problems and Strategies. In: *Ambic*, Vol. 14(6).

Shiva, V. (1989). *Staying Alive – Women, Ecology and Development*. Zed books. London, New Jersey.

Snodgrass, D.R. (1966). *Ceylon: An Export Economy in transition*. Homewood, Illinois, Richardd. D. Irwin, Inc.

Snyder, M. (1990). *Women: the key to ending hunger*. The Hunger Project, New York.

Spears, J.S. (1978). Wood as an energy source: The situation in the developing world. Paper presented at the 103rd annual meeting of the *American Forestry Association*. Washington D.C., The World Bank.

Srinvvasan, L. (1990). *Tools for community participants: a manual for training trainers in participatory techniques*. PROWWESS-UNDP. Technical Series.

Stephard, A. (1985). Social forestry in 1985: Lessons learned and topics to be addressed. ODA. Social forestry New Work Paper.

Swaminathan, S. (1982). Environment: Tree versus Man. In: *India International Center Quarterly*, Vol. 9, No. 3 & 4.

Tinker, I. (1987). The real rural energy crisis: women's time. In: *The Energy Journal*.

UNCED, (1992). *United Nations Conference on Environment and Development*, Press Summit of *Agenda 21*, Rio de Janeiro, Brazil, June 1992.

UNCED, (1992). *The global partnership for environment and development*. UNCED, Geneva, April 1992.

Warren, K.J. (1987). Feminism and ecology: making connection. *Environmental etnics*. Vol. 9, No. 1, pp. 3-20.

Whitmore, T.C. and Sayer, J.A. (eds.) 1992, *Tropical deforestation and species extinction*. Chapman and Hall, London.

WHO, (1984). *Biomass fuel combustion and health*. Geneva. Memeographed.

Wickramasinghe, A. (1986). *Agro-Ecological Land Potentials of the hill-country*. A paper resentation at the Institute of Fundamental studies, IFS. Kandy, Sri Lanka, October, 1986.

Wickramasinghe, A. (1988a). *Status of women in Rural Sri Lanka*. Women in Development (WID) Canadian International Development agency. CIDA, no. 12, Amarasekera Mawata, Colombo 5, Sri Lanka.

Wickramasinghe, A. (1988b). *Ecological impacts of Hill-country farming Systems*. The Natural Resource, Energy and Science Authority of Sri Lanka, Colombo, Sri Lanka. (Funded by NORAD).

Wickramasinghe A. (1988c). Impact of land-use practices on environmental conditions of the hill-country of Sri-Lanka. *Sri Lanka Journal of Social Science*, vol. 11, no. 1 & 2.

Wickramasinghe, A. (1990a). *Deforestation, Rural energy and women in forestry development in Sri Lanka*. A paper presentation at the Second national convention on women's studies. 20-22nd September, 1990. Centre for women's Research (CENWOR), Colombo, Sri Lanka.

Wickramasinghe, A. (1990b). Social and Economic Aspects Dealing with the Conservation and Management of Natural Forest in Sri Lanka. *In Harmony with Nature*, Proceedings of the International Conference on Conservation of tropical biodiversity. (eds) Yap Son Kheong and Lee Su Win, pp. 592-607, Kuala Lumpur, Malaysia.

Wickramasinghe, A. (1990c). *Use of wood energy for rural industries in Sri Lanka*. A paper presented at the Regional Expert Consultation on Wood Based Energy Systoms for Rural Industries, held in Hat-Yai, Thailand, 12-16 March 1990, FAO.

Wickramasinghe, A. (1990d). Farm and village forest and Land use practices in two villages in Sri Lanka. *MPTS Research Network Series*, No. 9, Forest/Fuelwood Research and Development Project (F/FRED).Bangkok, Thailand.

Wickramasinghe, A. (1990e). Social and economic factors affecting the use of tree crops the small scale farmers. *The proceedings of the National Network Meeting on Multipurpose Tree Species.* (22-24 in March 1990). H.P.M. Gunasena (editor). University of of Peradenya, Sri Lanka.

Wickramasinghe, Anoja. (1991a). *Gender issues in the management of homegardens: A case study of Kandyan homegardens in Sri Lanka.* A paper presented at the International Symposium on Man-Made community, integrated land-use and biodiversity in the tropics. 26-30 October, Xishuanbanna, Yunnan, The People's Republic of China.

Wickramasinghe, Anoja. (1991b). Artocarpus in Sri Lanka: rice tree of the rural poor. *Farm Forestry News Letter*, vol. 5, No. 1.

Wickramasinghe, Anoja. (1991c). *Women, energy and environment.* A paper presented at the workshop held at Agrarian Research and Training Institute (ARTI), Colombo, Sri Lanka.

Wickramasinghe, Anoja. (1991d). The user's perspective in selecting tree species for farming systems. *Multipurpose tree species in Sri Lanka:* Research and Development. Proceedings. 2nd Regional workshop on: Multipurpose tree species (MPTS), (ed) H.P.Genasena. University of Peradenya, Sri Lanka.

Wickramasinghe, Anoja. (1991e). *Gender analysis and Forestry.* A case study submitted to FAO, Rome (unpublished).

Wickramasinghe, Anoja. (1992a). Women and equity in forestry: A case study in Sri Lanka in *Sustainable and Effective Management system for community Forestry,*(eds) H. Wood and W.H.H. Mellink, pp. 91-105. RECOFTC, Bangkok, Thailand.

Wickramasinghe, Anoja. (1992b). Women equity and natural resource management, Occasional Working paper in *Women's Studies and Gender Relations*, Vol. 1, 1992. The Centre for Research in Women's Studies and Gender Relations, The University of British Columbia, 2206 East Mall, Vancouver B.C. V6T 1Z3, Canada.

Wickramasinghe, Anoja. (1992c). *Village agroforestry systems and tree use practices: A case study in Sri Lanka.* MPTS Network Research Series, No. 17. F/FRED Project, Bangkok, Thailand.

Wickramasinghe, Anoja. (1992d). Identifying tree-breeding objectives of the small scale farmers in Sri Lanka. *Research on Farmers' Objectives for tree Breeding* (eds). John B. Raintree and David A.

Taylor, pp. 18-24. Windrock International Institute for Agricultural Development, F/FRED, Bangkok, Thailand.

Wickramasinghe, Anoja. (1992e). *Forests in the lives of rural women, a case study on a forest fringe community.* A paper presented at the Conference on: Women, Environment and Development, organized by Ruk-Rekaganno, Environmental Foundation Ltd. 22 February, 50/7C, Siripa Road, Colombo, Sri Lanka.

Wickramasinghe, Anoja. (1993a). Gender-specific features in forest and tree uses in South and Southeast Asia. *MPTS Network Research Series,* No. 19, F/FRED, Bangkok, Thailand. p. 59.

Wickramasinghe, Anoja. (1993b). *Comments on Secretariat Note on Women and Forestry Session.* Presented at the 15th Asia-Pasific Forestry Commisstion, 9-13th August, Colombo, Sri Lanka.

Wickramasinghe, Anoja. (1993c). *Women in the management of Forests and tree resources in Asia.* Paper presented at Fourth Women in Asia Conference, 1-3 October, 1993, The University of Melbourne, Australia.

Wickramasinghe, Anoja. (1993d). *Women and Forests: the issues pertaining to the management of forest resources,* (waiting publication).

Wickramasinghe, Anoja. (1993e). *Gender issues in the management of Forest Resources.* Paper presented at the Conference on Forest Management and Sustainable Development, 4-8 October, Kandy, Sri Lanka.

Wickramasinghe, Anoja. (1993f). Women's Roles in Rural Sri Lanka, *Different Places, Different Voices: Gender and Development in Africa, Asia and Latin America,* (eds.) Janet Momsen & Vivian Kinnaird. pp. 159-175, Routledge: London.

Wickramasinghe, Anoja. (1993g). The different impacts of Deforestation and environmental Degradation on men and women: Issues for Social Justice. *Journal of Human Justice.* Vol. 5 No. 1. 1993 (in Press).

Wickramasinghe, Anoja. (1993h). *The vital linkage between Health and Environment: Examples from Sri Lanka.* Paper presented at the Conference on Health and Environment, at the 20th National Conference on International Health (NCIH), June 20-23, Arlington, Virginia, U.S.A.

Wickramasinghe, Anoja. (1993i). *The socio-economic and conservational issues attributing to the effective management of the*

Adams's Peak Wilderness. Forestry Planning Unit, Forestry department, Colombo, Sri Lanka.

Wickramasinghe, Anoja. (1993j). *Multiple use of Forest Trees by the Rural Communities.* A paper presented at the 4th Regional Workshop on Multipurpose trees. 12-14 March 1993, Kandy, Sri Lanka (Proceedings in Press).

Wiersum, K.F. (1990). *Planning agroforestry for sustainable land use.* pp. 18-33. In: Planning for agroforestry, W.W. Budd, I. Duchhart, L.H. Hardesty and F. Steiner (eds.). Elsevier Science Publishing Company. New York. U.S.A.

Wilson, E.O. (1988). *The current state of biological diversity, in Biodiversity* (eds. E.O. Wilson and F.M. Peter), National Academy Press, Washington D.C., pp. 3-18.

World Resources Institute and the International Institute for environment and Development. (1986). *World Resources* 1986, New York: Basic Books. U.S.A.